ANNA DI RONCO

POLICING ENVIRONMENTAL PROTEST

Power and Resistance in Pandemic Times

First published in Great Britain in 2023 by

Bristol University Press
University of Bristol
1–9 Old Park Hill
Bristol
BS2 8BB
UK
t: +44 (0)117 374 6645
e: bup-info@bristol.ac.uk

Details of international sales and distribution partners are available at
bristoluniversitypress.co.uk

© Bristol University Press 2023

British Library Cataloguing in Publication Data
A catalogue record for this book is available from the British Library

ISBN 978-1-5292-2875-5 hardcover
ISBN 978-1-5292-2876-2 ePub
ISBN 978-1-5292-2877-9 ePdf

The right of Anna Di Ronco to be identified as author of this work has been
asserted by her in accordance with the Copyright, Designs and Patents Act 1988.

All rights reserved: no part of this publication may be reproduced, stored in
a retrieval system or transmitted in any form or by any means, electronic,
mechanical, photocopying, recording or otherwise, without the prior permission
of Bristol University Press.

Every reasonable effort has been made to obtain permission to reproduce
copyrighted material. If, however, anyone knows of an oversight, please contact
the publisher.

The statements and opinions contained within this publication are solely those of
the author and not of the University of Bristol or Bristol University Press.

The University of Bristol and Bristol University Press disclaim responsibility
for any injury to persons or property resulting from any material published in
this publication.

Bristol University Press works to counter discrimination on grounds of gender,
race, disability, age and sexuality.

Cover design: blu inc
Front cover image: alamy/Bethany Sewald
Bristol University Press use environmentally responsible
print partners.
Printed and bound in Great Britain by CPI Group (UK) Ltd,
Croydon, CR0 4YY

To my *nonni,* as always

Contents

List of Images		vi
List of Abbreviations		vii
Notes on the Author		viii
Acknowledgements		ix
Introduction		1
one	Flexing the Muscles of Power: Policing Urban Eco-Justice Activism During the Pandemic	16
two	Power, Consumption, Disorder and Protest in Inner-City Centres	47
three	Atmospheres of Eco-Justice Resistance During the Pandemic	71
Conclusion		100
Notes		106
References		110
Index		131

List of Images

1.1	Police presence at a NOTAV protest in front of the city hall on 2 December 2020	30
1.2	Barriers to limit access to Trento's main square, Piazza Duomo, during an eco-justice protest on 5 June 2021	32
1.3	Police presiding over the geognostic survey site in Povo (Trento) on 9 February 2021	33
1.4	Local police car blocking access to the footpath on 13 February 2021	34
3.1	Stilt walker pulling yellow-and-black tape across road signs to encircle the protest area	84
3.2	Tape tied across a bar's outside umbrellas	84
3.3	Banner at the entrance of the San Martino neighbourhood	84
3.4	Large banner hung by activists on the fence of Scalo Filzi	88
3.5	The San Lorenzo bridge	92
3.6	'MISSING' wild bear sticker on a lamp post on San Lorenzo bridge	92
3.7	NOTAV flag on Doss hill as seen from San Lorenzo bridge	92

List of Abbreviations

ASBO	anti-social behaviour order
DIGOS	General Investigations and Special Operations Division
DPCM	Decree of the Italian Prime Minister
NGEU	NextGenerationEU funding scheme
NOTAP	anti-Trans-Adriatic Pipeline movement
NOTAV	anti-high-speed railway committee
PNRR	Piano Nazionale di Ripresa e Resilienza ('National Recovery and Resilience Plan')
PSPO	public spaces protection orders
SC	StopCasteller campaign
TAP	Trans-Adriatic Pipeline
TAV	*Treno Alta Velocità* ('high-speed railway')
XR	Extinction Rebellion

Notes on the Author

Anna Di Ronco is Senior Lecturer in Criminology at the Sociology Department of the University of Essex, UK, and Director of its Centre for Criminology. Her research focuses on the social control of incivilities, the local governance of sex work, urban resistance and criminalized eco-justice movements and their social media protests.

Acknowledgements

This work is indebted to eco-justice activists in Trento, who warmly welcomed me among them, offering an endless source of inspiration, critical knowledge and intellectual vitality during difficult and dark pandemic times. To them and all other oppressed eco-justice defenders, I dedicate this book.

I would like to thank my colleagues and dear friends Nigel South and Robin West, who provided feedback on early ideas around this book and have always been extremely supportive of my research ideas and projects. I am also deeply thankful to those who invited me to present my research in their institutions and offered feedback on it, including: Rossella Selmini and the critical criminological 'crowd' of the University of Bologna; and Ascensión (Chon) García Ruiz and Manuel Maroto from the Complutense University of Madrid. Part of this work was written during a visiting fellowship at the Oñati International Institute for the Sociology of Law (IISL) in the Basque Country, which offered a bucolic and intellectually stimulating environment where I could progress with my writing. This book was also written in other cities I hold dear: Trento, Colchester and Berlin.

Many friends supported me through this journey, and although I cannot mention them all here, I would like to especially thank Marta Di Ronco, Annalisa Santin and Fiamma Terenghi, who heard about this book more than others (and sometimes even joined me in eco-justice meetings and protests).

Last but not least, I would like to thank Davide for his care, love and relentless intellectual stimulus, which always dares me to think beyond the taken for granted.

Introduction

This book focuses on the policing and social control of eco-justice movements during the COVID-19 pandemic, as well as on activist practices of resistance in the same period. First identified in China in December 2019, the extremely contagious SARS-CoV-2 virus quickly spread worldwide, causing over 6 million deaths at the time of writing (WHO, 2022). Since then, and for at least three consecutive years, the COVID-19 pandemic has affected our lives in many different ways. For example, during the pandemic's first year and in periods with reported increased infection rates, children could not go to school, old people in elderly homes could not see their loved ones, families based in different countries could not reunite and many workers had to adjust to remote modes of working (at least the most fortunate ones who could do so). In addition, social gatherings were limited or even banned by many governments all over the world, with the implication that the right to protest was also often seriously curtailed in the interest of public health.

Since the first year of the pandemic, our knowledge of the virus certainly improved and led to the production of vaccines, which have been administered to billions of people worldwide. With higher levels of immunization reached among the population and decreasing infection rates, social life has slowly gone back to what it was before. In Europe right now, people who wish to organize a public protest can perfectly do so following the standard procedures that, in most cases, were in force before the pandemic began. This has not always been the case during the pandemic (and, for that matter, may not be in the future should other outbreaks occur or infection rates reach sky-high levels). Indeed, strict regulations affecting the right to protest were introduced in many European countries at several points in time to mitigate the spread of the pandemic, including during the winter of 2020–21, which is the time frame considered in this study. Regulations impinging on the right to protest often included temporary protest bans, caps on the number of people able to lawfully protest and requirements to maintain social distance or wear a mask during protests. In many Western countries, violations of these rules were strictly enforced by the police and mostly punished with administrative fines.

The right to protest has not only been limited for public health reasons. Indeed, both right before and during the COVID-19 pandemic, many countries including Italy (see, for example, Di Ronco, 2021b), Australia (Kurmelovs, 2021) and the UK (Gov.uk, 2022), enacted regulations clamping down on public protest. In the case of the UK, legislative changes on this matter were explicitly motivated by the need to entrust the police with greater powers to deal with allegedly new disruptive protest strategies adopted by eco-justice movements like Extinction Rebellion (XR) (see Chapter Two). Even when not meant specifically for eco-justice movements, punitive normative frameworks on public protesting are often enforced against them, as their mobilizing is gaining momentum and many high-profile protests are being organized in many countries all around the world. I will discuss the Italian and UK normative frameworks on protesting later in this book; for now, suffice to say that a more intense penalization of an ever-increasing eco-justice activism is at odds with the need – recognized by many governments, not least at the recent Conferences of the Parties (COP26 and COP27) – to fight climate change in an effort to prolong the life of our planet and avoid extinction. State actors' ambivalence towards eco-justice activists has also been observed by Hasler et al (2020: 12), who speak of a 'paradoxical' and 'multi-dimensional' relationship between the state and those who seek to protect the environment.

INTRODUCTION

In addition to punitive regulations, protest policing powers also seem to have intensified in recent years, especially as a result of a series of social, economic, political and health 'crises' (Burnett et al, 2022), and large and petty 'states of exception' (Fritsch and Kretschmann, 2021), which have allowed protest policing strategies to strengthen and become more visible, even spectacular at times. Among the recent crises, one would also need to include the COVID-19 pandemic – a health crisis also worth analysing for its implications for the policing and social control of public protest.

The existence of tough(er) policies and policing practices on public protest does not mean, however, that resistance is ultimately stifled or neutralized. Quite the opposite, activists often find new ways to keep their messages in public sight in spite of the many obstacles placed on their mobilizing by restrictive policies and strengthened policing powers. This is especially the case for eco-justice campaigners, whose numbers have likely risen in recent years due to growing concerns among the public (and young people in particular) around the climate emergency (Ares and Bolton, 2020). For example, during England's first national COVID-19 lockdown in 2020, people were required to stay at home and avoid meeting anyone outside their households (Brown et al, 2021). In spite of being unable to meet, activists continued their campaigning for climate and ecological justice, finding new ways to pass their message across. In Wivenhoe, the town where I live in Essex (UK), the local section of XR left a white stroller during the first COVID-19 lockdown containing a message in support of the Climate and Ecological Emergency (CEE) Bill, which is still being discussed in Parliament at the time of writing. At the same time, a number of graffiti, tags and flyers started to appear in Wivenhoe and nearby Colchester, including such captions as: 'let's not go back to business as usual', 'doomed 2020', 'grow a better future', 'act now' and 'eco emergency'. Eco-justice movements also used other ways to keep their grievances in public sight during the first stages of the pandemic, for example, through visual imagery circulated on social media (Ismangil and Lee, 2021) and digital strikes and protests (Finnegan, 2020). Other strategies discussed in this book include the use of stickers, stencils, graffiti, flags and flyers, in addition to the organization of flash mobs, in-person gatherings and protests all around the city. These visual and performative activist efforts allowed eco-justice movements not only to maintain their grievances in plain sight but also to disseminate ideas and imaginaries of alternative, more sustainable ways of living, which I will address later in the book.

This book focuses on two Italian eco-justice movements based in the northern city of Trento and pays attention to the dimensions of both power and resistance in urban struggles for eco-justice. It first examines how eco-justice activists were more intensively policed during a particular period of the pandemic health crisis (from December 2020 until September 2021), when protesting was either banned or allowed under strict conditions, which I shall review later. Second, it discusses eco-justice protesters' restricted access to, and invisibility in, the city centre, mostly through the existing multidisciplinary literature on the governance of consumption-focused urban centres. Third, it considers some of my autoethnographic experiences with eco-justice activist practices of resistance, which enabled me to discuss how the two movements succeeded in spreading their messages in spite of the obstacles placed on their gathering, mobilizing and protesting by COVID-19 and other regulations, while also offering some suggestions for future research in green critical sensory criminology.

Situating this book theoretically

This book contributes empirical data to the study of the policing and social control of urban eco-justice protest, while also focusing on activist practices of resistance and the progressive and transformative scenarios they often glimpse.

INTRODUCTION

Theoretically, this book is situated within the theoretical framework of critical criminology and specifically contributes to the field of *green critical criminology*. Generally speaking, green criminology is an approach that studies crimes and harms affecting the environment, the planet, human and non-human species, and environmental (in)justice (Brisman and South, 2020). The green criminological component of this book is evident from the type of social movements being addressed in it: eco-justice movements fighting for environmental, ecological and/or species justice (see later; see also White, 2013, 2014). However, green criminology is not necessarily always critical or attentive to the power dynamics and systems of oppression that breed ecological damage and harm; although grounded in critical criminology (Ugwudike, 2015; Sollund, 2017), this burgeoning perspective is open to multidisciplinary and interdisciplinary approaches (Ruggiero and South, 2013), as well as to various theoretical orientations (see, for example, White, 2013), some of which may be lacking critical depth. This book remains faithful to the critical origins of this perspective (see, for example, South, 2014; Stretesky and Lynch, 2014) and takes a critical-analytical stance towards the study of the policing and social control of eco-justice movements, and their resistant practices. In line with critical green criminology, this book also focuses on the concept of *harm* beyond legal and legalistic definitions of 'criminal harm', understanding it as any form of suffering caused by people in positions of power to the environment, ecosystems, non-human animals and plants, as well as other humans – with the latter also including the physical, emotional and economic harms suffered by eco-justice activists through intense policing, surveillance and criminalization.

The book also uses literatures that are grounded in (green) critical criminology, such as critical sensory criminology (McClanahan and South, 2020; see also Chapter Three), and criminological perspectives that are in close alliance to it, such as cultural criminology (see, in particular, Chapters

Two and Three). The fact that this book positions itself in, and contributes to, green critical criminology does not mean that it only draws on this body of knowledge to inform its critical analyses. In fact, it spans well beyond criminology and draws on other relevant multidisciplinary and interdisciplinary scholarships across the social sciences and humanities, including those addressing protest policing (in Chapter One), the urban regulation of disorder (in Chapter Two) and sensory methodologies in the study of atmospheres (in Chapter Three).

Ultimately, the book critically analyses the policing and social control of eco-justice activism during the COVID-19 pandemic, and tries to anticipate how they will unfold in its aftermath, while also drawing on previous literature on protest policing and the social control of disorder in the city. Hopefully, this effort will help problematize the current efforts of policy makers to regulate an ever-growing eco-justice activism in the wake of the climate crisis. At the same time, mainly drawing on critical sensory criminology, this book also considers eco-justice visual and performative activist practices in the city, and aims at emphasizing their importance and transformative potential. Although resistance may seem (to some at least) fleeting, questionable and ultimately ineffective in challenging the power structures that cause or facilitate eco-justice harms, it holds a powerful potential for change that should not go unrecognized.

Case study

This book draws on a case study of two eco-justice movements in the northern Italian city of Trento. With approximately 112,000 inhabitants and an assessed relatively high quality of life,[1] Trento is the political and economic capital of the autonomous province of Trento. The city is surrounded by mountains, and the Adige River passes alongside the centre and through the city.

INTRODUCTION

The two eco-justice movements considered for this study are the local anti-high-speed railway (NOTAV) committee and the animal rights campaign known as 'StopCasteller' (SC). As I will discuss later, both these movements were considered dangerous by the police well before the pandemic began.

Since 2013, the NOTAV committee in Trento has opposed the building of a high-speed railway and rail freight line passing through the city through an underground railway tunnel.[2] The railway is part of a broader European project to develop the so-called 'Trans-European Transport Network' (TEN-T), a network of (among others) railway lines, roads, ports, airports and railroad terminals that will help connect Europe and improve social and economic cohesion.[3] Within such a project, the existing Fortezza–Verona railway line passing through Trento is meant to be quadrupled.[4] The project is also part of a larger urban revitalisation plan for Trento that aims at making the city more 'sustainable ... green, modern and liveable' (Grottolo, 2021), with less car traffic, more walkable and green spaces, and new systems of public transport. However, this project has been challenged by eco-justice activists for different reasons, including the negative impacts it will likely have on humans, non-human animals, plants and the ecosystem. According to NOTAV activists, the project will, among other things, increase air pollution through disrupted car traffic, affect groundwater and further contaminate it through the dangerous chemicals already present in some of the areas along the envisaged tunnel route, and cause structural damage to the buildings above the tunnel (Coordinamento Trentino NOTAV, 2019).

Conflict around this project intensified during the pandemic, in particular, after the provincial government obtained funding for it through the Piano Nazionale di Ripresa e Resilienza ('National Recovery and Resilience Plan' [PNRR]).[5] The PNRR is a plan through which the Italian government detailed the ways in which it intends to use the NextGenerationEU (NGEU) – the economic recovery funding package rolled out by the European Union (EU) to support its member states to recover from the pandemic and help them prepare a better future for the next generation of Europeans.[6] The PNRR also includes the *Treno Alta Velocità* ('high-speed railway' [TAV]) in Trento, which is presented as part of the Verona–Brennero high-speed rail freight line connecting the north of Italy to Germany (PNRR, 2021). If the TAV project were to be implemented, as it seems likely, it will have to be completed by 2026 – the year when the NGEU fund for recovery and resilience will cease.

The study also considers another eco-justice movement: the animal rights campaign known as 'StopCasteller' (SC).[7] This campaign has been promoted since the beginning of the pandemic by the local social centre known as 'Bruno',[8] whose members have a long experience with police repression due to the many social and environmental campaigns they have promoted (for a general overview of Italian social centres, see Mudu, 2004). Indeed, along with other animal rights movements, SC activists within the Bruno social centre are not merely focused on animal liberation but also broadly campaign against multiple forms of animal and human oppression (see also Johnston and Johnston, 2017, 2020; Stephens-Griffin, 2022).

Since 2020, the SC campaign has opposed the local predator control policy specifically addressing wild bears, which envisages the 'incarceration' and even killing of the 'dangerous' ones that prey on livestock or come too close to human settlements. The SC campaign has not only challenged this regional policy but also exposed the poor conditions of some wild bears (including the bears known as M49, DJ3 and M57), which were held captive in a prison-like centre in the city's suburbs, known as 'Casteller' – a name that inspired the campaign's own name.[9]

Some conceptual clarity

This book uses the terms 'eco-justice' and 'resistance' consistently throughout its chapters. The term 'eco-justice' is borrowed from the green criminological literature and encompasses at least one of the following ideas of justice: environmental justice, ecological justice and species justice (White, 2013, 2014). Environmental justice addresses the negative impacts of environmental harms on humans at the intersection of race, gender, class and other systems of oppression. Ecological justice focuses on protecting specific environments (of which humans are also part), while species justice considers the protection of non-human animals and plants in particular. The two movements I consider in this book focus on at least one of these three ideas of justice: while the NOTAV Committee mainly fights against the TAV project and for environmental and ecological justice, the SC campaign mainly (though not only) champions species justice. In their narratives, these two movements also use these terms as translated in Italian ('*giustizia ambientale*', '*ecologica*' and '*specista*'). For its overarching reach (while also matching the many ideas of justice pursued by the two considered movements in their actions), in the book, I decided to use the term 'eco-justice' to refer to both the movements and their struggles.

In this book, I also use the term 'resistance' to explore activist practices. In particular, I use this term to refer to both visual activist practices (including graffiti, stickers, stencils and paste-ups) and performative activist practices (for example, street-based protests and flash mobs). I am aware, however, that the term 'resistance' is a contested one in the criminological literature, particularly, though not only, among scholars holding ultra-realist perspectives (see, for example, Winlow et al, 2015; for a critique within cultural criminology, see Hayward and Schuilenburg, 2014). These critical positions dismiss microlevel resistance, mainly for its alleged often unclear political intent (see Hayward and Schuilenburg, 2014), and some also argue for the need for organized and institutionalized political leadership to properly configure 'resistance' (see Winlow et al, 2015). Similar to Naegler (2021), I believe that these positions dismiss the transformative practices of social movements and the rich critical knowledge produced by them (see also Di Ronco and Chiaramonte, 2022). Along with Naegler, I ground my objections on in-depth empirical engagement with social movements; those like me who engage with social movements would not hesitate to recognize that such collectives often produce essential critical knowledge and that they test visions of social and eco-justice from below through practices of resistance (whatever we – academics – want to call them). Following Ferrell (2022), I also take the view that immediate, everyday acts of resistance, such as the gluing of a sticker on a lamp post, are, in fact, resistance *tout court* – even without proving the political intent of the people who stuck them there (an intent that is often present, as demonstrated by Stephens-Griffin [2022] in the case of minor or 'petty' events by animal rights movements). As I will demonstrate in Chapter Three of this book, everyday acts of resistance have the ability to convey powerful messages and provide a glimpse of better futures, which hopefully inspire and embolden further activism against power. As Ferrell (2022: 605) also argued: 'small acts of resistance hold the potential for blowing little holes in the daily practice of the social order – for opening lived spaces in which the taken-for-granted operations of social control can be challenged and problematized'. For all these reasons, I decided to use and embrace the term 'resistance' throughout the book.

Doing qualitative research during the pandemic: some reflections

For this research and book, I collected data from 1 November 2020 until 31 August 2021 in the city of Trento. During this period, teaching in my institution was moved to online platforms, and I was thus able to temporarily relocate to Italy, close to my family. Being in Trento gave me the opportunity to engage in an extensive ethnography on the policing of eco-justice movements and their resistance during a specific phase of the pandemic. Perhaps unsurprisingly, conducting research during the pandemic was not easy; it was particularly difficult during the first period of my fieldwork – between November 2020 and April 2021 – when Italian regulations to mitigate the spread of the pandemic were particularly tough. During this time, people in Italy could not leave their homes from 10 pm until 5 am, unless that was absolutely necessary for work or health reasons, or was otherwise needed. Movements across municipalities were prohibited on bank holidays (for example, between 25 and 6 December 2020, on 1 January 2021, and between 9 and 10 January 2021) and across different regions. In particular, travelling across regions was forbidden between 21 December 2020 and 15 January 2021 – a ban that was then extended by different Decrees of the Prime Minister (known as DPCMs) and remained in force for most regions until 25 April 2021. Starting from late February 2020, though with exceptions (for example, Easter), restrictions varied across regions according to the regional and most up-to-date epidemiological data (Governo Italiano, 2021, 2022).

During the initial part of my fieldwork and beyond, I obviously had to comply with the existing rules governing the pandemic – and so did activists, who could mostly meet up during the day and only organize protests under tight conditions and not at all on some specific days (for example, during the bank holidays in the winter break). From November 2020 until April 2021, therefore, my fieldwork mostly consisted of following the two eco-justice movements online through their social media accounts and attending their meetings online and offline (though only a few face-to-face meetings open to the public were organized during that time). In addition, I also used this time to walk around the city, taking pictures of markers of eco-justice visual resistance – an effort that inspired the reflections included in Chapter Three.

Regulations in Italy started to be relaxed only late in April 2021. It was then when the two considered movements resumed in-person protests and assemblies, including in the evenings. From April 2021 onwards, I could therefore engage with the movements more regularly, attending their weekly meetings and organized protests and events. In total, during my fieldwork, I attended 15 in-person events organized by the NOTAV Committee and the SC campaign, including weekly meetings, flash mobs and demonstrations, which I discuss later in this book. During all this time, the wearing of facial coverings was compulsory in Italy in all closed spaces and, at times, in open yet crowded spaces, particularly in periods when the epidemiological data generated concerns among local or national authorities.

Overall, the pandemic and its governance slowed down my access to the two considered eco-justice groups. With little face-to-face opportunities to meet activists, it took me more time to introduce myself as a researcher – and activist – and get closer to them. In general, as well as from my previous experience of studying eco-justice movements (see Di Ronco and Allen-Robertson, 2021), I can say that accessing open social movements can be relatively easy (depending on the attitude of the researcher) but also time-consuming: it literally takes time to access a group and be trusted by its members. Certainly, my credentials as a researcher (that is, my institutional web page) facilitated such a process. However, being trusted also means regularly attending meetings, engaging with the organized activities and helping with the organizing, when need be, and all this takes time. It is my belief that it is precisely because I engaged in such activities for a long period of time, which stretched well beyond the end of my fieldwork in September 2021, that I was able to collect such rich data on eco-justice activism and its policing during the COVID-19 pandemic.

Structure of the book

This book is based on three substantive chapters, followed by some concluding remarks. Chapter One analyses the policing of eco-justice protest during the pandemic. The chapter relies on a ten-month ethnography that involved attending relevant online and offline events, conducting interviews and focus groups with eco-justice activists, and having informal conversations with key police officers in charge of public order in the city. The findings point to the intensification of protest policing strategies during the pandemic, in particular, those predicated upon police visibility. The maximization of police visibility also went hand in hand with the increased invisibility of eco-justice demonstrators: they were made more invisible during their movements and were mostly displaced outside the historical centre. In the chapter, I analyse the findings through the relevant critical criminological literature on protest policing, the penalization of dissent and social control in consumption-focused urban spaces. I further examine the exclusion of protesters and others deemed disorderly individuals from consumption-focused urban centres in Chapter Two.

The displacement of eco-justice and other forms of protesting from urban centres is indeed discussed in depth in Chapter Two. The chapter specifically focuses on inner-city areas and outlines how their engrained emphasis on consumption informs dynamics of inclusion and exclusion in the urban space, as well as negatively affecting protesters and the right to protest. The chapter reviews the urban studies and critical criminological literatures, arguing that while some people are 'designed in' (Pali and Schuilenburg, 2020) to these spaces – and prompted to consume and behave well via subtle preconscious sensory and affective cues and nudges – others are 'designed out'. Among the latter, there are the urban poor, the homeless, young people, sex workers, migrants and protesters. In short, those who are 'designed out' tend to be individuals who are constructed as a hindrance to the realizing of the full economic potential of consumption-oriented inner-city areas. As the chapter illustrates, these subjects are pushed out of regenerated city centres via several means, including gentrification-led displacement, hostile architecture, technologies affecting the senses and incivility regulations and measures. The second half of the chapter specifically focuses on policies and practices that limited the right to protest in inner-city areas during the pandemic, with a view to safeguarding commercial values and promoting consumption. The chapter contends that the example of Trento is not unique; rather, other cities and countries in Europe and beyond have displaced protesting outside consumption-oriented urban centres. In such contexts, the right to protest has increasingly been understood as a hindrance to businesses and post-pandemic economic recovery.

In Chapter Three, I shift the focus from the policing and social control of eco-justice protesting to activist practices themselves, turning to their 'affective atmospheres' and the powerful messages and radical possibilities they often convey. In the chapter, atmospheres are defined as the qualities that shape what given places and events feel like and mean to people, also in light of their foreknowledge, memory, expectations and anticipation for the future (see Sumartojo and Pink, 2018). Using the flexible framework of Sumartojo and Pink (2018), in the chapter, I engage in three autoethnographic exercises that enable me to reflect on a number of atmospheres I encountered both during a NOTAV protest and while walking on a bridge and spotting markers of green visual resistance on and from it. Reflecting on my sensory and affective experiences of being *in* such 'atmospheres of resistance' enabled me to *know about* and *through* them; in particular, I was able to acknowledge the harms that humans cause to non-human animals and the environment, and to imagine a future devoid of such harms – and hence more respectful of the more-than-human surrounding us. The chapter contributes to the burgeoning field of 'critical sensory criminology' (McClanahan and South, 2020) and to the criminological literature on affective atmospheres, demonstrating the importance of studying atmospheres of resistance within green critical criminology – a study that, I will argue, should also involve activists and collaborations with scholars from other disciplines.

INTRODUCTION

The concluding chapter brings together and synthesizes the empirical and theoretical insights of the previous chapters. It concludes by identifying directions for future (green) critical-criminological research in the area and by discussing how green critical criminologists can support and enhance activist struggles for eco-justice.

ONE

Flexing the Muscles of Power: Policing Urban Eco-Justice Activism During the Pandemic

Introduction

Studies in critical and green criminology have highlighted how the repression of activism, and of eco-justice activism in particular, is increasing and becoming more pernicious all over the world (Maroto et al, 2019; Ruggiero, 2021a; Vegh Weiss, 2021a; Szalai, 2021). This literature has stressed the fact that state repression is often accompanied by the discrediting and depoliticizing of eco-justice movements, with the latter often being misrepresented as 'eco-terrorist' and ideological enemies impeding economic progress (Hasler et al, 2020). Attempts to discredit, delegitimize, depoliticize, silence and marginalize activists have been understood by Ferree (2004) as falling with the concept 'soft repression' – a non-violent form of repression that nonetheless has negative impacts on social movements and their mobilization. Indeed, as Jämte and Ellefsen (2020) suggested, when a social movement is labelled and stigmatized, activists tend to be less open about their involvement with it to avoid being outed and receive social sanctions; in turn, this has a negative impact on the mobilization of social movements, in particular, social movements that are more open and inclusive (see also Muncie, 2020).

After being framed as 'enemies' through soft repression, eco-justice movements are also often the target of hard repression (or 'over-criminalization') (see Vegh Weiss, 2021b) through violence, harassment, surveillance and criminalization. In some instances, activists are not just criminalized but even killed for their grievances and fighting. The killing of environmental activists is most common in the Global South, where in the year 2020 alone, 227 deaths were recorded, making it 'the most dangerous year on record for people defending their homes, land and livelihoods, and the ecosystems vital for biodiversity and the climate' (Global Witness, 2021: 10). Using the Environmental Justice Atlas, Scheidel et al (2020) suggested that killings, violence and criminalization tend to occur more often in mining and land conflicts, and where Indigenous groups are involved in protesting and mobilizing. In the case of Brazil, the killing of environmental activists has also been deemed to have been incited by its right-wing populist leaders (see Toledo et al, 2021; Bombardi and Porto Almeida, 2022).

In a recent collection of critical criminological essays, Ruggiero (2021a) examined how eco-justice activists become victims of what he called 'state–corporate terrorism'. Partially reworking Sykes' and Matza's neutralization techniques, and drawing from Merton's work on powerful offenders, he emphasizes the use by states and corporations of an economic reasoning to justify harmful corporate activities and, ultimately, activists' killing: corporate activities – against which activist often fight – are presented as crucial to the economic growth of the country and functional to the general well-being of the population. In a nutshell, as Whyte (2016) put it, corporate harmful misconduct is 'common sense' and hence generally acceptable precisely because corporations are perceived as socially beneficial.

In the Global North, when their lives are usually not at stake, environmental activists are often subject to intense surveillance. As mentioned earlier, activists are often framed as 'domestic extremists' or 'terrorists' (Salter, 2011). As a 'security threat', they are often strictly monitored by public and private actors (Walby and Molaghan, 2011; Lubbers, 2012; Ellefsen, 2018; Hasler et al, 2020; Brock and Goodey, 2022), undercover agents (Schlembach, 2018) and militarized police, especially during protests (Waddington, 2007; Wood, 2014; Passavant, 2021). The intrusive surveillance work often conducted on eco-justice activists has been well exemplified by Crosby and Monaghan (2018) in their work on the policing of Indigenous movements in Canada. In their account, the 'security state' – a conglomerate of national security agencies, police forces, private corporations and seemingly 'neutral' regulators – has come to define First Nations opposing extractive capitalism and asserting self-determination as security threats to be subject to tight surveillance through the collection of information in databanks and institutional partnerships with various private and public actors (see also Gobby and Everett, 2022). Another example also involving the surveillance of Indigenous peoples by both public and private actors is that of the protest against the Dakota Access Pipeline (DAPL) in North Dakota (US), which passes through the Standing Rock Indian reservation. Using leaked documents published by *The Intercept*, Hasler and colleagues (2020: 524–6) argue that a 'collusion' between law enforcers and private security firms succeeded in constructing non-violent NoDAPL protesters as 'terrorists', 'enemies' and a 'threat to society'. Such constructions legitimized 'the continued need for extraordinary security measures' (Brown et al, 2017, quoted in Hasler et al, 2020: 524), including the surveillance of activists during on-the-ground demonstrations, on social media and in their daily lives and movements.

As the next section will illustrate, in recent years, eco-justice activists have also been strictly controlled at demonstrations and rallies, as well as punished through both criminal penalties and administrative sanctions (Maroto et al, 2019). During 'states of exception' and recent political, economic and health 'crises', critical scholars have also observed the criminalization of much public protest (Fritsch and Kretschmann, 2021), the use of fines to regulate the right to protest (Martin, 2021, 2022) and the general strengthening of police powers against protesters (Lee, 2021; Martin, 2021, 2022; Burnett et al, 2022).

This chapter focuses on the policing of eco-justice protest during the COVID-19 pandemic and on activists' experiences and perceptions thereof. It draws on a ten-month ethnography in the city of Trento (Italy) and, more concretely, on interviews, focus groups and informal conversations with activists of two selected eco-justice movements, as well as informal conversations with key actors in charge of public order in the city. Ultimately, this chapter aims to illustrate what changed in the policing of eco-justice activism during the pandemic and explains these changes through critical scholarship on protest policing, the criminalization of dissent and critical accounts of public order in neoliberal inner-city spaces. The chapter is structured as follows. In the first section, I review the literature on protest policing and the repression of (eco-justice) activism in Western countries, focusing specifically on Italy, which is the country this book concentrates on. After providing an overview of the methods I used to collect and analyse the data, I present the three main themes that emerged from the analysis and discuss them through relevant critical criminological literature. The chapter concludes with a discussion of the findings and with some reflections on the possible extension of pandemic protest policing strategies to the pandemic aftermath.

Policing and penalization of protest in Western countries

Since the 1980s, in Italy as well as in other Western countries, the adopted protest policing style is that called the 'softer negotiated approach' – an approach that seeks containment through negotiation and cooperation with protesters, while avoiding physical confrontation and coercive intervention (della Porta and Reiter, 1998; Waddington, 2007).[1] Such an approach replaced one based on the incremental use of force against protesters, which was mainly used by the police prior to the mid-1970s. This changed in the late 1970s and 1980s, when – as put by Marx (1998: 255) – 'the velvet glove increasingly comes to replace, or at least cover, the iron fist' of the police.

However, the fact that the police tend to use the softer negotiated approach does not mean that their iron fist will not reveal itself in certain instances. Styles of protest policing can indeed become more confrontational, aggressive or coercive depending on the situation, the type of protesters being faced by the police and a number of other different factors, including: the organizational features of the police; the police culture; pressures from the public and the political sphere; and the police's own knowledge of their role and of the external reality (della Porta, 1998; della Porta and Reiter, 1998; della Porta and Fillieule, 2004; Waddington, 2007). The last factor of 'police knowledge' is of particular importance here, as it also includes stereotypes that the police have about certain groups of demonstrators who are ultimately perceived as 'bad', 'troublemakers', 'dangerous' or 'transgressive' – labels that are often attached to eco-justice movements (for the example of the British Stop Huntingdon Animal Cruelty, see Ellefsen, 2018). According to the relevant literature on protest policing (della Porta, 1998; della Porta and Reiter, 1998; Waddington, 2007; Gargiulo, 2015), while 'good' protesters are usually identified as workers or family men protesting to protect their jobs and support union demands, 'bad' protesters are often associated with leaderless and non-hierarchical groups who protest for 'confused' and 'abstract' issues that do not concern them directly and whose behaviour often appears unpredictable. Waddington (2007: 118) confirms this by arguing that: 'the iron fist is still likely to be revealed – and used, if need be – in situations involving transgressive (as opposed to contained) protesters and within political climates of opinion which vilify culturally or politically dissenting groups, their goals and their means of achieving them'. As Waddington (2007) suggests, perceived 'transgressive' protesters, such as anti-globalization movements, are usually dealt with through a large number of police personnel and containment measures that include the establishment of 'no-protest' security zones. della Porta (1998), who extensively analysed styles of protest policing in Italy, observed the great deployment of riot-geared officers already in the 1990s, in particular, to police protests organized by social centres and autonomous groups, which often embrace eco-justice struggles. Since the early 2000s, strategies of selective incapacitation based on the isolation of symbolic places (for example, through fences and police vehicles blocking access to places) have been employed against 'transgressive' anti-globalization demonstrations during international summits. In Italy, these strategies have also been used against domestic demonstrations, as della Porta and Zamponi (2013) demonstrated in their study on the 2011 anti-austerity mobilization in Rome. For Gillham and Noakes (2007), the establishment of extensive no-protest zones and the introduction of other practices aimed at temporarily incapacitating 'transgressive' protesters mark a new phase of protest policing called 'strategic incapacitation' – a new approach that often coexists with that of negotiation. This approach involves several practices, including the establishment of no-protest zones, the increased utilization of less-than-lethal weapons, the strategic use of arrests and a reinvigoration of the surveillance of 'transgressive' movements (for the example of the policing of anti-fracking protests in the UK, see, for example, Gilmore et al, 2020).

In Italy, many of the practices of strategic incapacitation have been used against the NOTAV and anti-Trans-Adriatic Pipeline (NOTAP) movements and their demonstrations. NOTAV and NOTAP are two eco-justice movements fighting against two megaprojects at the two geographical ends of Italy: while the NOTAV movement opposes the Turin–Lyon high-speed railway project (TAV) in the north-western Piedmont region, the NOTAP movement fights against the Trans-Adriatic Pipeline (TAP) in the south-eastern Puglia region. In both cases, these are 'strategic' and multi-million euro megaprojects partly funded by the EU (Di Ronco and Chiaramonte, 2022). For both NOTAV and NOTAP, the relevant literature documents the use of police violence during protests, the militarization of the relevant construction sites (which are protected through high fences, razor wire and 24/7 police surveillance), the surveillance of activists and the over-criminalization of activists through both administrative and criminal law (see, for example, Chiaramonte, 2019; Di Ronco et al, 2019; Di Ronco and Allen-Robertson, 2021; Tuzza, 2021; Di Ronco and Chiaramonte, 2022). Criminalization has so far been particularly severe for the NOTAV movement, which is 21 years older than the movement fighting against TAP: while the opposition against the TAV project dates back to the year 1989, the one against the TAP pipeline is much more recent, dating back to 2010 (Di Ronco and Chiaramonte, 2022). The longevity of the NOTAV movement explains the extent of its criminalization: so far, as Di Ronco and Chiaramonte (2022: 427) contend, it has 'involved over 1,500 people (as both suspects and defendants), about 150 criminal proceedings, the use of pre-trial preventive measures, and extremely serious criminal charges including that of terrorism' (for which all activists were however acquitted) (see also Chiaramonte, 2019).

Criminalization is not only used against 'transgressive' protest; according to Fritsch and Kretschmann (2021), in both petty and large states of exception, it is also employed against protests traditionally considered less problematic. Protesters – whether defined as 'transgressive' or not – are targeted not only through criminal law but also through administrative measures and fines. Fines are usually issued for two reasons: the commission of minor offences (contraventions or violations); and the adoption of behaviour considered a 'nuisance'. In Italy, when protest organizers fail to inform the authorities about their organized protest or to observe the conditions imposed on the protest by the police, they commit a minor offence (contravention) and can be fined for that.[2] Protesters, as well as other individuals deemed 'uncivil', can also be issued an administrative ban in Italy and thus be prohibited by the mayor from entering given city areas, including: train and bus stations; parks; the areas around schools, universities and museums; and areas attended by tourists (Selmini, 2020).[3] Bans are not uncommon sanctions for dissenters in Italy; they can also be imposed by the *questore* ('head of the police') on people deemed to be 'socially dangerous' and hence even before an actual crime has been committed (Prison Break Project, 2017).

Starting from the 2000s, the use of administrative fines and criminal penalties against protesters and political dissenters has also been observed in Spain (Maroto, 2017; Oliver and Urda, 2019; Selmini, 2020). Similar to Italy, administrative fines have been issued in Spain for the commission of such violations as the failure to notify the police of a planned march and (when it does not constitute crime proper) for showing disrespect to police agents during their duties (Selmini, 2020).[4] Activists can also be fined for the adoption of 'uncivil' behaviour, such as littering through the distribution of flyers, using overly loud megaphones during protests and camping in public areas (Maroto, 2017; Selmini, 2020).

As observed by Maroto et al (2019: 6), the use of criminal law *and* administrative fines against protesters tended to be common in the mobilization cycle of the 2010s, which was characterized by 'intense intolerance toward economic and social disruption'. In such times, as the authors contend, repression was embraced by the neoliberal state to deal with social mobilizations and 'contain emerging political alternatives', in this way, revealing the increasingly authoritarian face of power (Maroto et al, 2019: 13).

More administrative measures have been introduced during the COVID-19 pandemic. Indeed, during the pandemic and its 'politics of exception' (Fritsch and Kretschmann, 2021), many Western countries adopted new regulations governing the everyday, including the mundane acts of handwashing, walking on the street, clothing, shopping, travelling, socializing (Young, 2021) and, of course, protesting. These regulations have limited the right to protest in public places and allowed the police to mostly sanction protesters with administrative fines if found in violation of the new set of rules. The latter often included a cap on the number of people who were able to lawfully protest and requirements to maintain social distance or wear a mask. These restrictions have mostly been justified in the light of the need to protect public health, with public protests being considered as having the potential of becoming 'super-spreader events' (Martin, 2021: 224).

Although courts have upheld the right to protest in a few cases (on the case of Australia, see, for example, Martin, 2021, 2022; for the case of Colombia, see, for example, Morato, 2021), protesting has been significantly limited in the first two years of the pandemic. Critical criminologists attentive to power dynamics and repression have carefully analysed COVID-19-related regulations and policing practices affecting the right to protest, among other things (see, for example, Martin, 2021, 2022; Fatsis and Lamb, 2021; Burnett et al, 2022). However, empirical studies that focus on the policing of protest during the pandemic and on activists' experiences thereof are still quite scarce. Among others, one exception is the study by Lee (2021), who produced an autoethnographic account of participating in an XR protest in New South Wales (NSW, Australia) during the pandemic. In the study, the author discussed the perceived high police presence and surveillance at the protest: the number of deployed police officers matched that of protesters, and police followed activists closely to detect violations of COVID-19 NSW health regulations (which, at the time, allowed only groups of a maximum of 20 people to lawfully protest). This study aims to contribute empirical data to the study of the policing of eco-justice protest during the COVID-19 pandemic and of activists' experiences and perceptions thereof. This is important because patterns of surveillance and control of activists may not dissolve with the end of the pandemic but outlive the virus and affect (sometimes even further) the lives of eco-justice activists and their mobilizing.

Study background: governing protest in Italy during the pandemic

During the period considered for this study, COVID-19-related DPCMs regulated the right to protest, among other individual behaviours. In particular, such decrees only allowed 'static' demonstrations to take place, which are assemblies that are tied to a specific place and do not involve rallies or marches.[5] Protesters also had to comply with social-distancing rules, wear masks and observe the conditions imposed on protests by the local *questore* on a case-by-case basis. Any violation of these rules and conditions could be sanctioned administratively through arrest and fines (see Article 18 of Royal Decree No. 773 of 18 June 1931, also known as Testo Unico delle Leggi sulla Pubblica Sicurezza or TULPS).

Generally speaking, the two key local public order actors in Italy are the *questore* and the prefect. According to Law No. 121 of 1 April 1981, the prefect represents the national government at the local level and has administrative general competence for local public order, while the *questore* has a technical-operational role and responsibility over public order. These actors regularly exchange information, including through a so-called 'provincial committee for order and public safety', where other actors are also involved (including the mayor of the provincial capital, the governor and the local heads of the Carabinieri and Guardia di Finanza).

To evaluate the level of risk posed by a public event, the *questore* mostly relies on the work of the General Investigations and Special Operations Division (DIGOS), which is a political policing unit. Based in the provincial police headquarters (*questura*), DIGOS agents collect intelligence information on political movements and upcoming events, and communicate with the event organizers during rallies. After a demonstration is notified to the *questura* (a formal act required at least three days ahead of the event), its threat level is assessed and an operational plan for its management is agreed upon. Since the Law No. 121 of 1981, the *questore* can request the assistance of different forces in the management of public order, including the rapid reaction units of the police (or Squadra Mobile), Guardia di Finanza and Carabinieri.

Methodology

This chapter focuses on eco-justice activists' experiences and perceptions of policing and social control before and during the pandemic. As mentioned earlier in the introductory chapter, this book specifically focuses on the NOTAV Committee and the SC campaign, which are both based in the northern Italian city of Trento (for details on the two movements, see the introductory chapter).

The research relies on a ten-month ethnography in the city of Trento (from 1 November 2020 until 31 August 2021) and, concretely, on extensive field notes, interviews and conversations with activists, supplemented by photos taken during environmental protests and rallies, flyers and booklets distributed by NOTAV and SC activists at events, and their social media posts. Ethnographic work started as non-participant observation but changed into participant observation halfway through the project, when the national regulations set out to mitigate the spread of the pandemic were relaxed and I was able to regularly attend organized meetings, assemblies and protests (see the introductory chapter). From the end of April 2021, I got particularly close to the NOTAV Committee, actively partaking in most of its organized activities. My relationship with this movement and its members did not end with the end of my fieldwork but continues to the present day in the form of periodic engagement during the times when I return to the city.

During fieldwork, I participated in 15 in-person events (weekly meetings, flash mobs and demonstrations), where I had numerous informal conversations with activists.[6] I also attended eight online events, which were mostly held by NOTAV and SC in the period when social gatherings were restricted (January–April 2021).[7] During fieldwork, I also ran a focus group with three prominent NOTAV activists who had knowledge of, or experience with, repression (17 June 2021), and conducted an interview with one spokesperson of the SC campaign (24 July 2021). In the focus group and during the interview, I specifically asked activists to reflect on their experiences of police surveillance, control and repression before and during the pandemic. During the ethnography, I took extensive fieldnotes within 24 hours of events and, when considered appropriate (that is, where not perceived as intrusive) (see Natali and McClanahan, 2017), produced visual documentation of public protests and meetings, and of police surveillance at them.

During fieldwork, I also sought to interview the then *questore* and other officers in charge of public order in the city (including DIGOS and the Squadra Mobile of the police) to compare activists' experiences and perceptions with theirs. I succeeded in having *informal conversations* with three high-profile police officers who, at the time of the research, were or had just been in charge of public order in the city. Only in one case was I allowed by the respondent to record the conversation, and, for one officer, my request for an interview had to be 'centrally approved' (by the Italian Police Headquarters in Rome). In all three cases, moreover, officers made it clear that they only agreed to an informal conversation – not to an interview – where they could informally speak about general police operational practices for the governance of urban disorder in Italy. Despite the general nature of their answers, police officers provided interesting insights into their operating practices, which tend not to be well known by the Italian public, mostly due to the secrecy attitudes held by the police in Italy (see, for example, della Porta and Reiter, 2003; Gargiulo, 2015; Fabini and Sbraccia, 2021; Tuzza, 2021). These insights complement the analysis and are discussed in the following sections. Through thematic analysis of the material collected during fieldwork, I identified three main recurrent themes, which revolved around 'police visibility', the 'militarization of space' and 'COVID-19-related restrictions'.

Findings

Police visibility

According to activists, the governance of public protest during the pandemic relied on the maximization of police visibility. In particular, the police deployed a greater number of riot-geared officers and riot-control vehicles than they did before, and used drones and ranger raptor vehicles, which activists "had never seen before" (NOTAV focus group). Overall, police presence during the pandemic was described by activists as "shocking" (NOTAV focus group), "exaggerated" (NOTAV webinar, 30 March), "incredible", "disproportionate to the number of people present at the protest" and "unprecedented" (interview with SC activist); in essence, as they put it, it was a symbolic way of "flexing muscles" and being perceived to be in control (NOTAV focus group).

The deployment of a high number of police officers was also used as a method to police 'transgressive' protest in Italy and beyond before the COVID-19 pandemic. Indeed, well before the pandemic, della Porta (1998: 232) documented the use of 'massive and highly visible police presence' in Italy to discourage violence in protests involving social centres and autonomous groups. Waddington (2007), among others, also discussed the deployment of large numbers of police personnel to police 'transgressive' protest in Western countries. In Trento, however, activists only reported seeing extremely high numbers of police officers at protests during the pandemic: the police used to be much less visible before COVID-19, despite them always considering these two movements as potentially problematic and transgressive. Indeed, according to the police officers I spoke to, NOTAV and SC had been considered potentially dangerous groups well before the pandemic began: while many activists of the SC campaign (who are also members of the Bruno social centre) were known to the police for past occurrences, the NOTAV movement was believed to be joined at times by local anarchists, who also had a history with the police.[8]

If the police always considered these two groups as potentially problematic, why did they change their protest policing tactics during the pandemic? For two of the three police officers I informally interviewed, no real change was to be observed: as they suggested, the deployment of a high number of police agents at NOTAV and SC protests was needed to avoid clashes and minimize the risk of physical harm on both sides. Only one police officer admitted that police numbers at protests in Trento increased in recent years and explained this through the force's 'promotional incentives', rather than through the groups' alleged dangerousness. In their account, in order to get promoted, the *questore* and head of DIGOS need to move frequently across police headquarters in Italian cities, thus losing their knowledge of the territory and of its social movements. To get promoted, these actors also need to avoid 'making mistakes' while governing public order, and thus losing control of a protest that then results in damages or injuries. To avoid making mistakes, *questori* and heads of DIGOS often request the deployment of a high number of police officers – a number that usually tends to be slightly higher than those requested by their predecessors for similar events. According to this officer, these practices activate a vicious circle that leads to more and more agents deployed to manage protests, often unnecessarily.

The substantial presence of police at protests is not uncommon at eco-justice protests (see, for example, Muncie, 2020) and was also documented by Lee (2021) in his autoethnographic account of an XR protest in NSW (Australia) during the pandemic. It was also observed during my fieldwork: for example, at an event on 2 December 2020 in the historical centre of Trento, I observed around 50 NOTAV protesters peacefully protesting in front of the city hall, which was surrounded by at least 40 visible agents in riot gear and 19 riot-control vehicles (see Image 1.1). Activists later described police presence at this event as "shocking" and "utterly unnecessary": they were peacefully protesting outside the building of the local council, which was however empty at the time (the council met online due to the then COVID-19 regulations) (NOTAV focus group).

According to the interviewed SC activist, strategies centred around police visibility are intentionally used to construct dissenters as dangerous, sending the message that these movements are to be feared, and hence reinforce the need for a

Image 1.1: Police presence at a NOTAV protest in front of the city hall on 2 December 2020

Source: Anna Di Ronco

tight control of activists. This strategy appears to be quite typical in the protest policing of the mobilization cycle in the 2010s; indeed, as Maroto and colleagues (2019: 15) argued: '[s]eeing police (live or on television) identifying protesters or charging on a demonstration sends the message that there is something wrong about what is happening that justifies police intervention'.

During the time of this research, activists were constructed as dangerous not only symbolically through substantial police deployment but also discursively in media and public debates; for example, they were accused of threatening the province governor (SC) and the mayor (NOTAV) in public speeches and flyers, with the province governor being assigned police escort for his safety (Selva, 2021). All these accusations – which Ferree (2004) understood as 'soft repression' – were firmly rejected by activists and interpreted as deliberate strategies to undermine their grievances and delegitimize and depoliticize their fighting.

Militarization of space

Activists also noted another change affecting protest policing practices during the pandemic: according to them, the police made a greater use of material obstacles like barriers to regulate people's access to places, resulting in what they described as the "militarization" of public spaces (interview with SC; NOTAV focus group) (see Image 1.2). Containment measures, including no-protest zones, were also used to police 'transgressive' protest before the pandemic (Gillham and Noakes, 2007; Waddington, 2007), yet not with that intensity against eco-justice movements in Trento.

An example is provided by an SC event held on 10 April 2021, which saw the deployment of high numbers of police officers, along with police cordons and barriers, all around the city centre – and in front of the building hosting the provincial government, in particular. In that instance, barriers served the purpose of both isolating this key building and of narrowing the entrances to the main squares, thus limiting and controlling the number of people who could access all at once. As SC activists put it, the city centre then was "completely sealed off"[9] and 'militarised … as if we came here today to destroy [it]' (Global Project, 2021). This event also saw the police striking protesters with batons right in front of the building hosting the provincial government after activists attempted to place a net on the barriers (see Global Project, 2021). The net had a symbolic meaning: it symbolized the cages of the Casteller prison, where wild bears were held captive. The isolation of some symbolic places associated with political power through fences and police cordons is not uncommon in the policing of protest in Italy. della Porta and Zamponi (2013), for example, observed the strategy of 'selective incapacitation' in 2011 during the (domestic) anti-austerity mobilization in Rome. As they argued, such a strategy focuses on protecting some areas with institutional buildings from protesters, rather than on protecting the right to protest or the city.

Image 1.2: Barriers to limit access to Trento's main square, Piazza Duomo, during an eco-justice protest on 5 June 2021

Source: Anna Di Ronco

While most of the protests I observed took place in urban settings, there were some instances in which they were staged in the city's suburbs. In February 2021, for example, NOTAV protesters held a number of events in the suburb of Trento

Image 1.3: Police presiding over the geognostic survey site in Povo (Trento) on 9 February 2021

Source: Anna Di Ronco

known as Povo, where a geognostic survey for the TAV project was taking place on ground that also hosted a vineyard. To protect the survey site from protesters, the police closed one of the two exits of the nearby train station to limit the walkable routes to it to only one; they also closely monitored that route and the area around the site through police cordons and substantial police presence, both during the day and at night. For example, on 9 February 2021, late in the evening, I observed agents of three forces patrolling the area; these included Guardia di Finanza, Carabinieri, Polizia and plainclothes DIGOS agents (see Image 1.3). At the intersection between the survey site and the public footpath, I also observed eight riot-control vehicles, with one of them shining a beam light to illuminate the site and temporary flash-blind all people walking on the path, thus weaponizing light in the service of police power (see Shalhoub-Kevorkian, 2017; Garcia Ruiz and South, 2019; McClanahan

Image 1.4: Local police car blocking access to the footpath on 13 February 2021

Source: Anna Di Ronco

and South, 2020). During that evening, less than 30 activists met in a close-by square to discuss the TAV project.

In the following days, the police also limited access to the footpath that runs close to the survey site to residents only.

For example, after an event organized by NOTAV activists on February 13, I was warned by a local police officer that I would get cautioned if I had not avoided the survey site area by taking a much longer walking path to the city, which I eventually took (see Image 1.4).

NOTAV activists defined the area of the survey site as 'militarized' territory, involving the deployment of an 'exaggerated' number of riot-geared agents and barriers (Coordinamento Trentino NOTAV, 2021; CSA Bruno, 2021a). During the focus group, NOTAV activists also reported seeing "shocking" increased levels of police presence in the city during this time: as they argued, police agents stayed at a hotel in the city and occupied many public parking slots with their vehicles and rangers. This attracted the public's attention (for example, people were seen taking pictures of the parked vehicles), as such a level of continuous police presence was understood as exceptional.

NOTAV activists understood the strategies of protecting the survey site in Povo as part of a police 'protocol' – a set of actions the police put in place when it comes to defending strategic construction sites from dissenters across the country. According to them, such a protocol involves the militarization of the area surrounding sites through fences and the 24/7 presence of the police; it also involves the exclusion of people from accessing the area (unless they are residents or have special police permissions) and the repression and criminalization of dissenters. Although this 'protocol' has not yet been implemented in its full force in Trento, the NOTAV Committee was aware of it through the experiences of the NOTAV movement in the Susa Valley and the NOTAP movement in Puglia mentioned earlier, with which it has tight connections and regular contacts. Indeed, both in the Susa Valley and in Puglia, the construction sites of the TAV and TAP megaprojects have been militarized through fences, razor wire, surveillance cameras, police and private security, as well as defended through the repression and criminalization of activists (see earlier; see also Chiaramonte,

2019; Di Ronco and Allen-Robertson, 2021; Di Ronco and Chiaramonte, 2022).

According to one of the police officers I informally interviewed, current strategies to protect the survey site are based not on a national protocol but, rather, on a local 'model'. Such a model was informed by a series of lessons learnt from the past (including a 2015 incident where activists were able to damage the geognostic survey drill in Trento)[10] and by one of the key officers' professional experience in Turin in the 1990s, where they learnt how to manage NOTAV protests against the first geognostic surveys in the Susa Valley. According to this officer, such a model in Trento involves the use of barriers to limit access to the survey site and the deployment of large numbers of police agents – with the latter strategy serving the purpose of keeping the site safe and avoiding clashes with protesters. As the officer put it when referring to the defence of the survey site in Povo: "No one was hurt, there were no clashes. ... It was a large operation ... using many resources. ... We are fully satisfied."

COVID-19-related restrictions

During fieldwork, the theme of the negative impact of the new COVID-19-related regulations on activist fighting also emerged. In particular, activists discussed the rules that prohibited people from moving from one municipality to another, or from one region to another, as well as night curfews. According to them, such regulations made it more difficult for activists to meet in person, limited in-person gatherings in the evenings and overall reduced their possibilities to discuss key issues, organize new campaigns and actions, and mobilize (interview with SC). For activists, these regulations also served the purpose of normalizing the absence of dissent in the city and hence making the public used to little to no urban protesting.

Activists also discussed the national regulations that, during the pandemic, only allowed protesting in static forms (that is, not involving parades). One example of a static protest in Trento is one organized by the NOTAV Committee on 15 May in the neighbourhood of San Martino – an area just outside the historical centre that will likely be affected by the TAV project. The event involved a number of static protests in different parts of the neighbourhood, where activists were able to discuss various project-related harms and place-specific impacts. Protesters moved from one area to the other by walking; however, to be able to comply with the relevant COVID-19 regulations allowing only static demonstrations, they had to wrap up their flags and hide placards and banners (they could unwrap flags and show banners only when standing still). I attended that protest and observed the organizers making people aware of this rule after being warned by DIGOS agents. Activists also mentioned other examples. For example, the SC activist I interviewed spoke about the protest organized by the SC campaign on 10 April 2021. In that instance, SC was only allowed by the police to hold static protests in two city areas outside the historical centre; however, similar to NOTAV in the preceding example, it was prohibited from holding flags, banners and posters while moving from one area to the other; it was also prohibited from walking through the centre.

Eco-justice protesting was also regulated at the local level by the *questore*, who could put conditions on protesting as they saw fit (and still can). As activists argued, demonstrations from SC and NOTAV in Trento were mostly allowed by the *questura* outside the historical centre. This was confirmed by one police officer I informally interviewed, who argued that the then *questore* had a preference for protests to be held outside the centre (and off its main square, Piazza Duomo, in particular). During the pandemic, however, not all protests were displaced by the *questura* outside the centre. For example, rallies organized by people opposing the COVID-19 vaccines (or 'anti-vaxxers') could be held in the centre, as well as, for a long time, protests opposing the so-called 'COVID pass' – a digital or paper certificate proving that its holder has been vaccinated, tested negative or recovered from COVID-19.[11]

Only early in November 2021 was the requirement of static protests outside the historical centre in Trento applied to anti-COVID-pass demonstrations. As one of the local newspapers reported, this was the result of complaints by local businesses, which lamented their decreased revenues on Saturday afternoons due to the frequent anti-COVID-pass parades in the city centre (Pontalti, 2021). Shortly after, similar grievances led the minister of the interior to extend this prohibition to all protests in Italy (Interno, 2021). In particular, the minister gave prefects the power to identify 'sensible urban areas, of particular interest to community life', where the right to protest could substantially be limited during the pandemic (Interno, 2021: 2).

The special protection of city centres and their economic value is not new in Italy. As mentioned earlier in this chapter, areas featured by the presence of, among other things, museums, archaeological sites, monuments, cultural heritage sites and sites attended by tourists – which, in Italy, are often located within city centres – can be subject to a sort of enhanced protection by the mayor. The latter, in particular, can decide to ban all individuals whose behaviour is thought to 'impair the access and use' of city areas by others (Di Ronco, 2018), including the urban poor, migrants, sex workers and even protesters (Selmini, 2020). It is worth mentioning that since 2008, local authorities can also request the presence of the military to patrol city centres in conjunction with the state and municipal police (Di Ronco and Sergi, 2019; Selmini, 2020) – a possibility that has since been used by many Italian cities (Selmini, 2020).

During my fieldwork, only two eco-justice protests were held in the historical centre: while the first was authorized by the police, the second was held in spite of the denied police permission. The first demonstration was organized by the NOTAV Committee on 2 December 2020 and the second by the SC campaign and other groups on 5 June 2021. They were both presided over by high numbers of police personnel and riot-control vehicles – a presence that was described by

NOTAV activists as "shocking" and "utterly unnecessary" (NOTAV focus group). I observed the second protest by SC during fieldwork. At around 2.30 pm, approximately 50 activists marched to the main square in the city centre, where they were expected by more than 30 visible riot-geared agents and 11 police cars (three of which were riot-control vehicles). After reading their messages out loud through a loudspeaker and distributing some flyers to passers-by, protesters were escorted out of the centre via secondary roads by the police. In this instance, activists knew of course that defying an order of the police could lead to sanctions and penalties. As they suggested, however, sanctions or even criminal charges are not necessarily issued immediately by the police; rather, they could come unexpectedly at any point in time, including right before large events and protests, as a way to discourage participation in them (interview with SC activist). Indeed, I observed the 'belated' charging of activists during my fieldwork: in December 2021, seven SC activists were accused of damaging the fence surrounding the Casteller 'prison' (where three bears were held captive) during a protest that was held more than a year prior (on 18 October 2020) (CSA Bruno, 2021b). Activists also identified other tactics used by the police to intimidate them, including threatening to arrest people at public protests (interview with SC activist), approaching people directly to try to discourage them from being part of the movement (NOTAV focus group) and stopping and identifying them during the regular distribution of flyers (NOTAV focus group).

Flexing the muscles of power in the policing of eco-justice protest during the pandemic

Eco-justice activists in Trento reported higher levels of militarized police presence at their protests during the pandemic and noted that barriers were used by the police with more intensity that in pre-pandemic times, with the effect of 'militarizing' public spaces and restricting and tightly controlling people's access to them. Heightened police visibility was matched by the increased invisibility of eco-justice protesters, who became less visible in the historical centre and in their movements during events (that is, activists could not wave flags and other protest symbols while walking).

High levels of police deployment and containment measures have also been used to police 'transgressive' or 'bad' protesters before the pandemic (della Porta, 1998; Waddington, 2007; della Porta and Zamponi, 2013), including to defend deemed strategic construction sites (for the cases of NOTAV and NOTAP, see, for example, Di Ronco and Chiaramonte, 2022). However, they had not been used against NOTAV and SC protests in Trento in spite of them being labelled as 'transgressive' by the police long before the pandemic began.

In Trento, intensified protest policing strategies during the pandemic can be explained in many ways, including through the promotional incentives seemingly existing within the force. As one of the informally interviewed officers suggested, promotion criteria tend to encourage the *questore* and head of DIGOS to request the deployment of ever-increasing numbers of officers to police protests in order to avoid "making mistakes" while governing public order and compromising their careers. Beyond this, the intensification of protest policing strategies can probably also be linked to the pandemic itself, which, as in other recent 'crises' and 'states of exception', involved the strengthening of police powers against protesters (see, for example, Fritsch and Kretschmann, 2021; Lee, 2021; Martin, 2021, 2022; Burnett et al, 2022), including through public displays of power against them.

Moreover, as Fritsch and Kretschmann (2021) remind us, public displays of police power against protesters also help construct activists as 'enemies' and escalate the responses against them – thus justifying the 'exception' itself. Indeed, as activists themselves suggested during fieldwork, high militarized police presence at protests operates at a symbolic level: it constructs protesters as a 'threat', depoliticizes their fighting and legitimizes police presence and methods. This has also been contended in the relevant literature (see, for example, Ferree, 2004; Maroto et al, 2019; Hasler et al, 2020; Jämte and Ellefsen, 2020; Fritsch and Kretschmann, 2021). Manning (2003), for example, noted how the practice of policing has an inherent quality of spectacle – a quality that often serves to communicate and mark moral boundaries, and reaffirm social order and state authority. Policing practices often communicate dramatic messages, and this is also the case here: highly visual policing strategies offer a spectacle of public order to passers-by, with a large number of officers geared up for the occasion, accompanied by highly visually impacting riot-control vehicles – a symbolic armoury deployed to reaffirm state authority, construct protesters as a 'problem' and send the message that police presence (and their eventual use of force and criminalization) is absolutely necessary to protect the public and maintain public order. The iron fist shines under the velvet glove, reminding protesters and the public of its existence and potential for violence.

As illustrated in this chapter, eco-justice protesting in Trento during the pandemic was also affected by both national and local COVID-19-related policies and practices. For example, activists viewed national and locally imposed curfews, as well as rules restricting people's mobility, as an impairment on their campaigning and mobilizing. They also discussed the negative implications of the national requirement to only hold static protests in the city: in Trento, this meant that when demonstrators wanted to move from one place to another, they had to hide their paraphernalia to avoid being seen as part of a parade or a procession, which was prohibited under the national decrees tackling the COVID-19 pandemic.

Eco-justice protesting was also limited by local policing practices; in particular, activists were prevented from organizing protests in the historical centre by the local *questura*, which mostly allowed demonstrations to be held in areas outside the centre. Indeed, during the period considered for this study, only one eco-justice demonstration was allowed to be held in the historical centre. This demonstration was organized by NOTAV on 2 December 2020 and was presided over by high numbers of police agents and vehicles. A parade was also organized in the historical centre by SC on 5 June 2021; however, this was done in spite of the denied police permission and ended with demonstrators being escorted out by the police through secondary roads – to avoid being seen by the public. It is worth emphasizing that during the time of the research, demonstrations around other issues (for example, against the COVID-19 vaccines and pass) could be held in the centre.

The displacement of eco-justice protesting outside the historical centre certainly had to do with the 'transgressive' and 'dangerous' labels the police attached to the two groups under study: SC and NOTAV were both constructed as dangerous by the police and were thus more tightly controlled and regulated than other groups. In addition, let us not forget that the city centre is a repository of key economic and commercial interests, which can be seen as endangered by 'transgressive' and 'disruptive' forms of protesting. Early in November 2021, precisely these considerations led the *questura* in Trento to extend the ban on holding protests in the centre to 'disruptive' protests against the COVID pass. As local newspapers reported (Pontalti, 2021), this was the result of the pressures put on the police by local businesses, which were 'bearing the brunt of the economic losses linked to the protests that have been held weekly, for months now, against the green certificate [or COVID pass] and, in part, against [COVID-19] vaccines'. Later, in November, the minister of the interior further extended these protest bans by giving the power to prefects across Italy to identify urban spaces where protesting could be banned altogether. This directive was also motivated by the need to protect economic and commercial interests from disruptive demonstrations (Interno, 2021).

I will discuss the exclusionary dynamics generated in economically and commercially valued inner-city spaces in more depth in Chapter Two. Suffice here to say that in recent years, these spaces have increasingly come to be dominated by the mantra of consumption, which, in turn, fundamentally shapes the dynamics of inclusion/exclusion in such spaces. In essence, city centres have become 'container spaces', which – to put it in the words of Hayward (2012: 454) – are 'new scattered zones of safety and control, where otherness, irrationality and dissent are banished beyond boundaries of exclusion and distinction'. Borrowing the concept of 'urban container' from de Jong and Schuilenburg (2006), Hayward (2012) mostly applied it to private spaces and public spaces dominated by private interests, like inner-city spaces often are. Indeed, city centres are often hyper-protected spaces for the middle classes, predicated upon the exclusion of unwanted 'others', whose presence disrupts the interests of capital. Whether the urban poor, migrants, sex workers or eco-justice demonstrators, these 'others' are all accused of undermining the neoliberal consumption logic and economic interests, and are thus pushed out of city centres and other 'container spaces' alike. It is within this logic that we can read and interpret the local practice of allowing eco-justice protests mostly outside the polished and touristy centre of Trento: their 'transgressive' and 'disruptive' protesting is deemed to undermine commercial and economic activities and challenge power, and is therefore limited and tightly controlled.

Finally, in line with Maroto and colleagues (2019), this study found indications of the use by the police of both administrative measures and criminal law to control eco-justice protest (for Italy, see also Prison Break Project, 2017): while the police used administrative powers when imposing conditions on demonstrations (for example, including its location in the city), they employed criminal law through, for example, 'belatedly' charging SC activists with criminal damage (they were only charged a year after the occurrence of the alleged damages). The study also found that not all protests were equally policed in Trento during the considered time frame, with only eco-justice demonstrations being initially displaced outside the city centre. However, it also observed that this practice of displacement later applied to other types of demonstrations, including those opposing the COVID pass. These findings therefore support the argument by Fritsch and Kretschmann (2021) that during petty and large states of emergency, much protest is marked as hostile and criminalized – or, as in this case, punitively regulated. To be noted here is that criminal law could become an even more important punitive tool against SC and NOTAV in the future, as has happened to other eco-justice movements in Italy (for the cases of NOTAP and NOTAV, see Di Ronco and Chiaramonte, 2022).

What the future holds?

This chapter has addressed the policing of two eco-justice movements in Trento during the COVID-19 pandemic. Its findings have revealed the use of protest policing strategies predicated upon police visibility and spatial containment. Although such strategies were used by the police against 'transgressive' protest even before the pandemic, they had never been used with that intensity against the two eco-justice movements in question, despite them being considered problematic by the Trento police long before the pandemic began. In this chapter, I have explained the use of intensified police strategies through the incentives existing within the force, which seem to encourage those in charge of public order to deploy increasing numbers of police officers to manage public protests. These findings are also in line with recent critical-criminological literature which posits that during 'crises' and 'states of exception', police powers against protesters are usually strengthened (Fritsch and Kretschmann, 2021; Burnett et al, 2022), while also involving public displays of power against demonstrators. Such public displays of police power also help construct protesters as a 'problem' and depoliticize their protesting, while, at the same time, legitimizing police presence and their possible intervention (Fritsch and Kretschmann, 2021).

The chapter also found that increased police visibility was matched by an increasing invisibility of eco-justice activists: they were made less visible in the historical centre and during protests' movements (when they had to hide all protest symbols). The displacement of eco-justice protesters – and later also other types of demonstrators – outside the centre is a new development, though not a surprising one: it can be read and understood through the need to protect economic and commercial values enshrined in inner-city spaces from seemingly 'transgressive' and 'disruptive' protests.

This chapter has also emphasized the point made in the recent literature that authoritarian state strategies against social movements and their mobilizing rely not only on criminal law but increasingly also on administrative measures (Maroto et al, 2019). In particular, this chapter flagged the use of administrative powers by the *questura* to concentrate eco-justice protesting outside the historical centre and more tightly control it, mainly for economic (to protect commerce) and political reasons (to delegitimize eco-justice protesting). As illustrated earlier in this chapter, violations of the conditions placed by the *questura* on protests are mostly sanctioned through fines, which are also recognized a typical punitive tool against protesters during the pandemic (see Martin, 2021, 2022).

It remains to be seen if these protest policing strategies will outlive the pandemic and be applied beyond the current state of exception. One interviewed police officer in Trento seemed to hint at this when discussing the promotional incentives existing within the force: on their account, more and more officers will be deployed at future protests by those in charge of public order to protect their careers and minimize the risk that public protests (even the most peaceful ones) turn into public disorder. Important in this sense is also the analysis by Fritsch and Kretschmann (2021), who link the recent 'politics of exception' to the social construction of most public protesters as 'enemies' and their subsequent criminalization. Crucially, the effects of this labelling process do not seem to end with the state of exception; rather, for these authors, it is the othering of protesters that also legitimizes their criminalization after the crisis. As discussed previously in this chapter, some of the protest policing strategies used by the Trento police during the pandemic contribute to constructing eco-justice and other demonstrators as a 'problem'. Hence, following the argument by Fritsch and Kretschmann (2021), more intense policing, surveillance, exclusion and even criminalization of activists could also be potentially expected in Trento in the aftermath of the pandemic.

TWO

Power, Consumption, Disorder and Protest in Inner-City Centres

Introduction

This chapter focuses on power dynamics shaping how neoliberal inner-city space is regulated, experienced and lived by people, including by protesters – and eco-justice demonstrators among them. It is not by chance that this chapter focuses particularly on inner-city areas among other zones in the city. Inner-city space is, indeed, a 'special place' within the urban, at least in terms of the social construction of deviance and of social control. Over the past 40 years, in the UK and many European cities, such areas (or often sections of them) have frequently been subject to redevelopment or regeneration programmes, which have succeeded in revamping their look and image. Architectural and design changes have contributed to moulding or, in some cases, merely consolidating the consumption ethos that often governs these spaces. In other words, inner-city spaces have increasingly become what Keith Hayward (2004) called 'spaces of consumption and pleasure': spaces that offer various shopping, consumption and entertainment opportunities to people, tourists, party-goers and various other 'sensation-gatherers' (Bauman, 1997).

As this chapter will illustrate, an ingrained emphasis on consumption has had an effect on how inner-city areas are governed and, in turn, experienced and lived in by people. In its first part, this chapter focuses on how space has actively been implicated in the shaping of individual behaviour, including through engineered subconscious clues and nudges that push people into consuming more than what they had originally intended or into behaving in ways that do not discourage others from consuming as much as they wish to.

Inner-city spaces have been regulated through subconscious sensory and affective cues, as well as through more explicit means, for example, policies and practices that penalize certain 'unwanted' behaviour and bodies, while safeguarding these spaces' consumption ethos and commercial values. Prescriptive and narrow understandings of space are, however, not always unconditionally accepted by everyone in the city. As the second part of the chapter will illustrate, inner-city spaces often become a battleground where different constructions of space come into conflict – a conflict that often results in the criminalization of the less powerful, as critical and cultural criminological scholarship has cogently demonstrated. Drawing on studies from human geography, urban studies, sociology and criminology, the third part of the chapter tries to capture the ways through which people – including protesters – are excluded from regenerated inner-city areas beyond outright criminalization: displacement, hostile architecture, technologies affecting the senses and incivility regulations reframing people's (often entirely legitimate) behaviour as a nuisance or anti-social. In its last section, the chapter focuses specifically on how the consumption ethos of inner-city areas affected the right to protest during the COVID-19 pandemic. This last section mainly draws on this book's case study, as well as on recent examples taken from the UK and Europe, where protesting has increasingly been understood as a hindrance to business and limited via bans and stricter regulations.

Regenerated city centres: consumption and the subtle doings of power

In the contemporary globalized context, cities tend to compete against each other to attract and retain private capital – a phenomenon that Harvey (1989) understood as 'urban entrepreneurialism'. Cities are most successful at it – and at producing revenues – when they can boast regenerated spaces of consumption, leisure and cultural production catering to the needs of the wealthier, those holding consumption potential and the 'creative class' (Florida, 2002). These spaces are often located in inner-city areas or city centres, which tend to be 'shop windows' and the symbols of the identity of the city (Bergamaschi et al, 2014).[1] In such spaces, priority is given to private business interests and to making the privileged feel safe while shopping or generally engaging in consumption practices, often in privatized and closed 'sanitized' spaces like shopping malls (Davis, 1990; Hayward, 2004). In such closed and privatized spaces, conspicuous consumption is often incentivized through unconscious sensory and affective clues engineered through careful lighting, colour differentiation and sound frequency (Thrift, 2008; Hayward, 2012; Garcia Ruiz and South, 2019; McClanahan and South, 2020). Indeed, consumers may be affectively encouraged to consume – even compulsively – through subtle design strategies that are able to 'elicit physiological and psychological traits conducive to consumption' (Kindynis, 2021: 620). According to Kindynis (2021: 633), 'persuasion architectures' in consumer environments, such as shopping malls and other retail spaces, casinos and gambling environments, and night-time venues, preconsciously draw visitors to 'impulsive forms of consumption: shopping; gambling; or drinking further than they had originally intended'. In essence, as Kindynis (2021) suggests, the design of consumer environments can be a biopolitical strategy used to entice consumer behaviour, even in its most pathological and harmful forms.

Individual behaviour can be manipulated through subtle psychological triggers – or 'psychopolitical techniques', as Pali and Schuilenburg (2020: 6) put it – not only in private spaces of consumption but also in public consumption-focused spaces. Particularly important in this regard is the study by Schuilenburg and Peeters (2018), who focused on the 'De-escalate project' implemented in the 'smart' inner-city entertainment district of Eindhoven (The Netherlands). The project relied on the manipulation of the collective mood through the use of smart lighting technology, smell and sound design, which helped to reduce aggression, anti-social behaviour, noise nuisance, rubbish on the street and so forth. As the authors argued, the main objective of the project was not to exclude people from urban areas but, rather, to include them. Drawing on Foucault's series of ideas on pastoral power, Schuilenburg and Peeters (2018: 6) argued that pursuing security in this pastoral way is 'protective, inclusive, preventative, and caring'. Such a strategy is inclusive because it is instrumental to realizing the full potential of consumption-focused places like night-time districts; indeed, when people are nudged to 'properly' behave through subtle and unconscious psychological triggers, the area becomes enjoyable and pleasant for everyone, and this ultimately 'stimulate[s] an efficient, safe and consumption-focused use of space' (Schuilenburg and Peeters, 2018: 5). Even in such cases, however, exclusion is always lurking around the corner, still operating against: people who ignore the subtle nudges and engage in aggressive behaviour; those who do not consume in socially prescribed ways (see, for example, Measham and Brain, 2005; Hayward and Yar, 2006); and even people who are identified as a threat through predictive smart technologies (Pali and Schuilenburg, 2020).

In essence, various architectural and biopolitical techniques can encourage consumer behaviour in regenerated consumption-focused districts. They can do so in different ways, either by fomenting consumption, even in its more harmful manifestations, or by relaxing the atmosphere, enabling everyone to enjoy and consume without being confronted with physical or social disorder. The idea underlying these studies is that the architecture and design of urban spaces encode messages that can consciously or unconsciously be 'read' by people, telling them how to 'properly' behave and make a 'good' use of space.

Of relevance here is the work of Michel Foucault (1977), which has been used by geographers to examine how social order and discipline is instilled in space (see Holloway and Hubbard, 2001). In his work, Foucault did not discuss power relations in public spaces, however; rather, he was mainly interested in institutional spaces (such as the prison and the asylum), where control is exercised through surveillance and segregation in panopticon-like buildings. However, he also noted that in society, control is dispersed through complex networks of power and knowledge, and institutionalized in state programmes of education, welfare and social policy. These programmes, in particular, allow for disciplinary power to be internalized by people, turning them into docile, meek, 'good', productive citizens able to self-discipline and self-manage – in closed institutional spaces as elsewhere. Geographers (later also followed by criminologists) have widely drawn on these ideas to investigate power relations based on control and discipline in everyday spaces, not only in private and institutional settings but also in public spaces. Indeed, as Holloway and Hubbard (2001: 187) remind us, 'all places are potentially implicated in the reproduction of social orders' and should therefore be critically examined.

To conclude, a critical analysis of urban consumption-focused districts is important to reveal how affective and atmospheric control over people is exercised in a way that steers and shapes their behaviour. However, as the next section will illustrate, it must be noted that urban spaces and their atmospheres are open to adaptations that may escape the designers' intentions (Sumartojo and Pink, 2018; Popovski and Young, 2022) and often lead to contestation and resistance.

Regenerated inner-city spaces as a battleground

As mentioned earlier, in public regenerated and consumption-focused spaces, power can be exercised (and discipline instilled) in subtle ways, for example, through architectural features and design. More generally, urban planners and designers often encode the values of the dominant cultural groups – not least that of consumption – into the urban regenerated landscape, ultimately shaping how people sensorily and affectively experience urban spaces and behave in them. Processes of inscribing values into the geographical landscape are relevant to understanding the dynamics of inclusion/exclusion in the urban space: certain bodies, types of people, behaviour and embodied practices may be welcome and (culturally and commercially) valued in certain areas, while others may instead be deemed unacceptable or 'out of place'. Such processes are not straightforward: urban spaces often become a battleground, where the interests, lifestyles, values, norms and expectations of social groups clash against each other – with those of the wealthier usually prevailing over those of the others (Millie, 2011; Kiely and Swirak, 2022). This conflict is well captured by the theoretical work of Henri Lefebvre (1991), who argued that space is socially produced or constructed: as a product and medium of social relations, space is a site of negotiation, contestation and (re)appropriation. Following Lefebvre, contestation happens when authoritative constructions of space (as conceived by, for example, urban planners) are disrupted by the direct and embodied lived experience and symbolic spatial practices with which people engage in space.

Cultural and critical criminologists have also well captured the terms of this urban conflict, both theoretically and empirically. The cultural criminologist Jeff Ferrell (1997: 22), for example, conceptualized this conflict through the concept of 'cultural space', which he defined as 'those arenas in which young people and others construct meaning, perception, and identity'. Cultural space is indeed often contested, and this battle frequently triggers punitive sanctions of the powerless by the powerful. As Ferrell (1997: 22–3) argues in relation to youth subcultures:

> Within relationships of power, inequality, and marginalization, the control of cultural space is contested: while powerful adults attempt to define

and impose cultural space, less powerful young people attempt to unravel this imposition, to carve out their own spaces for shaping identity and taking some control over everyday life. ... Alternative cultural spaces that stray too far from zones of mainstream meaning, or that by their very existence breach the boundaries of accepted spaces, regularly elicit aggressive sanction ... including the clout of civil and criminal law.

In essence, many cultural spaces (plural) may coexist, for example, when different groups have conflicting understandings over the uses of public space. Yet, not all these cultural spaces may be equally tolerated by groups with political capital; as Ferrell suggests, in some instances, clashes over cultural space may lead powerful adults to impose their meanings, values, expectations and cultural norms onto young people via punitive sanctions.

Ferrell's work inspired a wealth of cultural and critical criminological studies that investigated conflicts between the cultural norms, values and expectations of different societal groups in urban settings – and, most notably, in consumption-focused places. Alone or with colleagues, Andrew Millie studied context-specific behavioural expectations leading to criminalization (Millie, 2011), and various instances where the aesthetic order of the city is challenged through the aesthetic practices of subcultural groups (Millie, 2017, 2019; Dickinson et al, 2022). In one of his 2011 articles, for example, Millie (2011) linked criminalization to the behavioural expectations that people holding political capital have in specific contexts. Using the example of the criminalization of graffiti, panhandling, street posters and billboards in Toronto (Canada), Millie suggested that behavioural expectations are informed by a combination of value judgements, which can be aesthetic, economic, prudential (dependent on considerations around the 'quality of life') and moral. These judgements, however combined, shape assessments of whether a given conduct is celebrated, tolerated, censured or even criminalized in public spaces. In his more recent work (Millie, 2017, 2019, 2022), as well as with other colleagues (Dickinson et al, 2022), Millie has explored the other side of the coin: no longer why powerful groups impose their expectations onto others through punitive interventions but, rather, the aesthetic practices that challenge mainstream aesthetic and social orders. These practices are those of urban interventionism (Millie, 2017), yarn bombing (Millie, 2019), street skateboarding (Dickinson et al, 2022) and guerrilla gardening (Millie, 2022) – all of which have the capacity to (at least temporarily) change not only the look but also the feel of cities.

A similar argument has been made by the cultural criminologist Theo Kindynis (2018: 511, emphasis in original), who used Lefebvre's theoretical work to conceptualize graffiti writing as a spatial practice that '*disrupt*[s] authoritative spatial orderings, while *superimposing* its own alternative social geography onto the city'. In other words, graffiti challenge dominant expectations around the looks and uses of space, and generate new layers of meaning and alternative imaginaries around it. Craftivism (McGovern, 2019) and music can also do that. For Fatsis (2019: 453–4, emphasis in original), for example, criminalized UK grime artists:

> use, draw on and 'produce' public space (Lefebvre, 1991) by 'spray[ing]' their lyrics like 'sonic graffiti' (Bramwell, 2015a: 11, 51) around the city in parks, on public transport and in neighbourhood corners. Such use of public space through beats (rhythms) and rhymes (lyrics) constitutes a broad and diffuse *agora* of sorts where meaning and culture are experienced as lived, embodied entities rather than as abstractions; creating opportunities for assembling citizens through speaking and listening (Oswell, 2009: 12).

In essence, airing criminalized grime music also challenges dominant constructions of space; in addition, as Fatsis (2019) suggested, it also generates spaces for meaningful political engagement in the city, particularly for those who feel and often are politically marginalized, in addition 'to being intensively policed and criminalized. As Dickinson et al (2022) contended, different aesthetic appreciations in and of space contribute to enriching our affective and sensory engagement with it and ultimately add value to city life; this should be appreciated and treasured by cities and its dwellers – not disapproved of and rejected (see also Bannister et al, 2006; Peršak and Di Ronco, 2018; Sendra and Sennett, 2020).

To be noted here is that aesthetic visual practices (such as street art), as well as performative practices (such as urban protest), can also be co-opted and commodified by cities for profit, as illustrated by the cultural criminologist Laura Naegler (2012) in her work on the gentrified neighbourhood of Sternshanze in Hamburg. In that neighbourhood, criminalized resistance was successfully incorporated into gentrification and commodified; seen as a proof of the 'authenticity' of the area, resistance was depoliticized and turned into a mere attraction for tourists. The topic of the commodification of resistance has also been investigated by urban scholars who have analysed the exploitation of urban art and creativity as a vector of further gentrification through different branding strategies (see, for example, Andron, 2018). In essence, urban resistance can be co-opted and depoliticized to serve the commercial and economic interests of power, particularly in areas that are being regenerated and gentrified. In the following section, I shall elaborate further on the strategies that are used by powerful groups to maximize their economic interests in regenerated areas. In particular, I shall focus on the ones that are predicated on the exclusion of the powerless.

Regeneration and exclusion

The regeneration of consumption-focused inner-city areas is often predicated on the exclusion of the 'disassembled others' who are negatively affected by neoliberal predatory formations and assemblages (Arenas and Sweet, 2019a): the urban poor, sex workers, migrants and, often, dissenters. There are different ways in which these 'others' are pushed out of regenerated spaces, including through incivility regulations, 'zero-tolerance' styles of policing (Bannister et al, 2006; Lundsteen and Fernández González, 2021) and even subtler measures – all of which have the effect of displacing, rather than resolving, social problems (Atkinson, 2003).

Gentrification-led displacement

Urban regeneration often involves so-called 'gentrification' – a process through which, at different degrees, the built environment and the socio-economic composition of given neighbourhoods witness a change. Although there are local specificities and its outcomes are varied (see, for example, Chabrol et al, 2022), gentrification tends to exacerbate socio-economic inequalities and further exclude the more marginalized and those perceived a 'risk' or a nuisance (Smith, 1996; Peršak and Di Ronco, 2018; Kiely and Swirak, 2022).

Whatever its characteristics in each given context where it unfolds, gentrification is thought to have one specific defining feature: it causes displacement (Elliott-Cooper et al, 2020; see also Davidson and Lees, 2010). As Elliott-Cooper and colleagues (2020) illustrated in their study, displacement can well be direct and short-term, for example, when residents (including social movements and squatter communities) are evicted from their homes by developers on occasions of large sporting events or large-scale developments. However, it can also be indirect and long-term, and thus be accomplished relatively slowly via a series of events: a neighbourhood is targeted by developers, retail changes, property prices gradually increase and many residents decide to leave because the cost of living has become unaffordable or the area no longer feels like home. As Elliott-Cooper et al (2020: 502) argued: 'displacement is never a oneoff event but a series of attritional micro-events that unfold over time, generating different emotions and mental states for those affected: anxiety, hope, confusion, fear, dislocation, loss, anticipation, dread and so on'. Indeed, even when displacement is not abrupt and forced but, rather, slow and pernicious, it can generate detrimental effects on people, including by inflicting mental and physical harms on them. Let us not forget that these harms are not evenly distributed across residents: most of the negative effects are usually felt by the working class, women and minority ethnic groups, among others. For them in particular, though also for other displaced people, displacement is a 'form of violence that removes the sense of belonging to a community or home-space' (Elliott-Cooper et al, 2020: 503).

Hostile architecture

In addition to displacing the powerless from gentrifying and gentrified spaces, planners and developers – backed up by local administrators – can also use other strategies to exclude the 'other' from valued consumption-oriented spaces. Such strategies often go unnoticed by visitors and inattentive well-off city dwellers. An example is that of urban design and architecture aimed at designing certain people out of specific consumption-oriented spaces. This is what so-called 'hostile' or 'defensive architecture' sets out to do: it involves the installation of street furniture, or the protection of the existing street furniture and of buildings, in a way that discourages some people from loitering or sleeping rough, while dictating narrow ways in which space should be used (see, for example, Davis, 1990; Bergamaschi et al, 2014; Peršak, 2021). Examples of hostile architecture include: ground and wall spikes, fences, studs or metal bars and rods in places where the homeless may seek shelter; benches with armrests or with slanted, convex surfaces, deterring the homeless from lying down; anti-skate handrails (with spikes or other metal elements on them); and 'skatestoppers' (for this last element, see

Dickinson et al, 2022). All these elements are aimed at removing some people from sight, in particular, those who do not have enough consumption potential to 'deserve' inclusion, such as the homeless and, often, young people. Defensive architecture not only excludes (and discriminates against) the urban poor and the young; it also makes the city more unpleasant, hostile and unfriendly to other bodies, including 'the overweight, the disabled, the pregnant, the ill' (Peršak, 2021: 68). For example, benches that are convex, slanted or with seat dividers are not only pushing away the homeless; they are also designed to only accommodate certain bodies, and often only for a limited amount of time. As Peršak (2021: 66) suggested, this sends the message that 'those who always seem to matter, who count as valuable citizens or city residents, are mobile, restless, working and paying consumers' – it is only to these people that urban spaces cater.

Technologies affecting the senses

Exclusion of unwanted people from spaces of consumption can also be achieved in other subtle yet equally 'anti-social' ways, including through technologies affecting the senses. Previously in this chapter, I addressed these technologies mainly via the work of Schuilenburg and Peeters (2018), who discussed how 'proper' behaviour in public spaces can be induced through non-physical elements of the built environment, such as light, smell and sound, which are captured by people through the senses. These technologies can be employed not only to include people in certain city spaces, as in the case studied by Schuilenburg and Peeters (2018), but also to exclude them. Examples of exclusion through technologies affecting people's hearing include the playing of loud classical music to disperse groups of young people (Hayward, 2012; Garcia Ruiz and South, 2019) and the so-called 'Mosquito device', which emits high-frequency noise that – when heard – hurts people (including by causing nausea and dizziness) and motivates them to leave. The Mosquito device was initially targeted at young people (Hayward, 2012) but was later redesigned to make noise audible to people of all ages (Little, 2015). The device has mostly been used against young people congregating and loitering in public consumption-oriented spaces (Little, 2015); however, there are reports of it also being used against older people, and the homeless in particular (see Kiely and Swirak, 2022). Less subtle technologies but equally affecting the senses while attempting to remove people from spaces are 'long range acoustic devices' (LRADs) (Garcia Ruiz and South, 2019; Brisman et al, 2021). Such devices can be used by the police in public spaces to disperse crowds and can have various harmful effect on protesters and demonstrators, including causing ear pain, dizziness, disorientation and even hearing loss and brain injuries.

Incivility regulations and measures

Another way through which the powerless are excluded from consumption-focused areas is through policies and associated policing practices against incivilities (nuisance, anti-social behaviour or disorder). Since the 1990s, many countries in Europe have adopted regulations targeting anti-social or disorderly behaviour, and aimed at increasing the perceived quality of life of the better-off (Peršak, 2017a). Such regulations have often been inspired by Wilson and Kelling's (1982) 'broken windows' thesis, which postulates that if incivilities are left unchecked, they will trigger insecurity and fear, breed further disorder, and lead to criminality spreading in the neighbourhood (Crawford and Evans, 2017). More broadly, these regulations have been informed by a 'pre-crime' logic (Zedner, 2007), or by the need to anticipate and govern 'risks', including those posed by disorder, which has increasingly been understood as a precursor to crime (Crawford, 2009; Di Ronco, 2016; Crawford and Evans, 2017; Pleysier, 2017). Incivility regulations have circulated through flows of policy transfer between countries and even cities (Selmini and Crawford, 2017), and have specifically been applied to regenerated inner-city areas.

In the UK and many continental European countries (Peršak, 2017a), regulations against incivilities have included individualized behavioural orders or bans, curfews for young people, and general local regulations prohibiting 'uncivil' behaviour in certain designated city spaces. These measures have been regarded by Peršak (2017b) as an example of 'criminalization through the back door', as they involve punitive regulation of human behaviour not through criminal law but, rather, through civil (in the UK) or administrative (in Europe) law, which are fields of law that are devoid of the legal safeguards typical of criminal law.

Criminal law is, however, not completely left out of the picture; in some instances, when civil or administrative measures are breached, the full force of criminal law is triggered. One of the most notable examples of this dual mechanism is provided by anti-social behaviour orders (ASBOs) in England and Wales. ASBOs were introduced by the New Labour government through the Crime and Disorder Act 1998 and within their broad 'respect agenda' (Burney, 2005; Squires, 2008; Crawford and Evans, 2017). By regulating behaviour that 'caused or was likely to cause harassment, alarm or distress' to the public or sections of it, ASBOs 'prohibit[ed] the defendant from doing anything described in the order' (Sections 1[1] and [4]). To be noted here is that what was considered as behaviour causing 'harassment, alarm or distress' was largely left to the discretion of the relevant local authorities, which could make a request for the issuance of an order to magistrates', county or sheriff courts. Particularly relevant for our purposes is that the breach of the prescriptions contained in the order immediately triggered the criminal law: the individual would face imprisonment[2] or a fine, or both (Section 1[10]). The criminal law philosophers Simester and von Hirsch (2006) described the ASBO system as based on a 'two-step prohibition', involving a civil-law order in the first step and criminal law in the second after the civil-law order has been breached. As these two authors noted, this system is particularly problematic when used to punitively regulate behaviour that is entirely legitimate, such as that of young people hanging about in public spaces (Simester and von Hirsch, 2006). This system has also been criticized by other criminal justice and criminology scholars, including Ashworth (2004), Burney (2008), Crawford (2009) and Ashworth and Zedner (2014), who noted how ASBOs circumvented and eroded traditional criminal justice principles, and violated human rights. These objections were not ignored by courts; quite the opposite, courts often upheld appeals by defendants and found ASBOs to excessively interfere with their fundamental rights (Burney, 2008; Di Ronco and Peršak, 2014).

ASBOs were finally repealed in 2014 by the Anti-Social Behaviour, Crime and Policing Act, though leaving an 'enduring legacy' of 'hybrid, and semi-criminal enforcement powers' that continued to be exercised well after the end of ASBOs (Squires and Stephen, 2010: 29). Indeed, as Squires and Stephen (2010) argued, problematic measures against anti-social behaviour entailing a form of pre-emptive or precautionary criminalization have continued to be adopted in the UK and have included dispersal orders, control orders and civil gang injunctions.

Worth mentioning here is the case of public spaces protection orders (PSPOs), which were introduced under Chapter 2 of the Anti-social Behaviour, Crime and Policing Act 2014. These orders can be used by local authorities to ban behaviour in certain designated spaces that 'has had, or is likely to have, a detrimental effect on the quality of life' of those in the locality. Breaching a prohibition or a requirement contained in a PSPO is a criminal offence punished with a level-three fine (£1,000), with the possibility of using fixed-penalty notices in place of prosecution (Brown, 2017).

There are at least two elements that are shared by both ASBOs and PSPOs. First, in both cases, the relevant legislation uses vague concepts that are open to subjective interpretations and thus eventual abuses. In the case of ASBOs, the vague concepts were those of 'harassment, alarm and distress', while for PSPOs, it is the notion of 'quality of life'. Second, breaching the prohibitions contained in both types of civil orders is subsumed in the remit of criminal law, thus disproportionately undermining civil liberties in public spaces (Brown, 2017). As reported by Manifesto Club, an organization that challenges 'the hyper-regulation of public spaces',[3] PSPOs have mostly been used in city centres (or in sections of them) against the homeless, young people and buskers.[4] Moreover, as Kiely and Swirak (2022: 94) suggested, in the case of PSPOs (and like measures) against the homeless: '[o]ne can observe ... a clear predominance and protection of commercial and private interests over the interests of marginal populations'.

Incivility measures that – when breached – trigger the criminal law are no prerogative of England and Wales; they also exist in Europe. In Italy, for example, access to economically valued urban areas can be restricted to certain deemed uncivil individuals through so-called 'urban DASPOs' (the urban version of the Divieto di Accedere alle manifestazioni SPOrtive, or ban on accessing football stadiums for deemed dangerous supporters). DASPOs are usually 48-hour bans that can be imposed by the mayor on individuals whose behaviour 'hinders the access and use' of public spaces and services by others; in such cases, individuals are also administratively sanctioned through fines of €100–300.[5] Mostly, these bans have been applied to homeless and drunken people, hawkers, unlicensed car park attendants, and sex workers – many of whom are migrants (Borlizzi, 2022). Areas that can be protected through DASPOs include: train and bus stations; parks; areas around schools, universities and museums; and places attended by tourists (Selmini, 2020) – with the latter in Italy often being situated in historical centres. Further, when the banned behaviour is repeated and there is a risk to 'public security', the *questore* can ban the individual from accessing certain designated areas for up to one year. It is the breach of the ban of the *questore* that triggers criminal sanctions: the individual faces arrest from six months up to one year.[6] Similar to the civil orders in England and Wales described earlier, in this case, applications of DASPOs can also be varied (and quite arbitrary), as they rely on subjective interpretations of vague concepts (for example, of behaviour hindering access and use of space, and of behaviour posing a threat to public security). In addition, breaches of the bans issued by the *questore* are also associated with criminal sanctions.

As mentioned previously, both in the UK and in Europe, incivility regulations have mostly been used to protect regenerated inner-city areas, where economic and commercial interests tend to be most prominent. In such areas, as Bannister, Fyfe and Kearns (2006: 923) have argued, 'anyone deemed to compromise the strict ethics of consumerist citizenship risks exclusion'.[7] While some administrative measures have specifically targeted these areas and their surroundings, others have been related to a broader urban perimeter but have de facto mainly been enforced in inner-city spaces. I have observed this in my own research on the regulation of sex work in the cities of Antwerp (Belgium) and Catania (Italy) (Di Ronco, 2021a, 2022). In Catania, for example, the 2017 incivility ordinance (which remained in force only until 2018) only targeted on-street sex workers and clients with administrative fines in the streets of the city centre (Comune Catania, 2017). In Antwerp, by contrast, the local *politiecodex*[8] prohibits on-street sex work across the whole urban territory. The latter, however, it is mostly enforced in the city centre, where on-street sex workers and their clients are thought to be present (Di Ronco, 2021a, 2022). More generally, in both cases, my research revealed how the 'problem' of sex work was constructed differently by local authorities and the police (as well as some third-sector associations) according to the space where it was deemed to be present: on-street sex work was mainly considered a nuisance undermining 'respectable' people's quality of life in city centres, while mostly associated with sex trafficking and exploitation in all other city areas. In other words, the 'problem' of sex work tended to be equated with serious crime committed by organized crime groups against vulnerable victims (women in particular) except when it took place in areas that are ingrained with the moral and aesthetic standards of 'respectable' citizens, and that enshrine economic and commercial values that are crucial to the city's wealth.

In Europe, administrative regulations and measures against incivilities have been used in inner-city areas to target not only sex workers (see also Villacampa, 2017) but also other 'unwanted' people: migrants (Crocitti and Selmini, 2017; Lundsteen and Fernández González, 2021), the homeless (Podoletz, 2017), young people (Gayet-Viaud, 2017) and protesters (Maroto, 2017; Selmini, 2020). As mentioned earlier, Maroto (2017) discussed how protesters in Spain have been targeted not only through criminal law but also through administrative offences and 'nuisance' ordinances. Through the latter in particular, activists have been fined for the adoption of 'uncivil' behaviour, such as littering through the distribution of flyers, using overly loud megaphones during protests and camping in public areas (Maroto, 2017; Selmini, 2020).

Restricting the right to protest in inner-city spaces

As illustrated earlier, protesters have been pushed out of regenerated and gentrified inner-city areas in several ways. For example, demonstrators have been targeted through technologies affecting or, better, hurting the senses – such as hearing – including through LRADs used by the police to disperse crowds. They have also been punitively sanctioned through incivilities regulations, with some of their activities being considered a 'nuisance' (for example, fly tipping, speaking too loudly through megaphones during protests and so on). However, that is not the end of the story. Recently, and most notably during the COVID-19 pandemic, protesting has increasingly come under attack in England, Europe and elsewhere in an effort to protect inner-city areas and their economic and commercial value.

An example of the social construction of protesting as a threat to businesses is provided by the UK Police, Crime, Sentencing and Courts Act 2022, which was first introduced in Parliament on 9 March 2021 (for the first and following versions of the bill, see UK Parliament, 2022b). Among other things, the act expanded police powers to put restrictions on processions, assemblies and even one-person protests, including on their start and finish times, route, size, and noise levels. Indeed, pursuant to the act, the police can now put restrictions on noise levels when the noise generated by people taking part in protests is deemed to cause a 'serious disruption to the activities of an organisation' in the vicinity (Gov.uk, 2022). The act also did much more: it lowered the fault element for offences related to breaching of conditions put on protests; it increased the maximum sentence for obstructing a highway; it penalized obstructing vehicular passage in the area around the Houses of Parliament; and it converted the common law offence of public nuisance[9] into a statutory offence with a maximum penalty of ten years' imprisonment (Gov.uk, 2022). As the explanatory notes to its first version suggested, the bill was meant to address the new disruptive protest strategies adopted in recent years by protesters: 'Recent changes in the tactics employed by certain protesters, for example gluing themselves to buildings or vehicles, blocking bridges or otherwise obstructing access to buildings such as the Palace of Westminster and newspaper printing works, have highlighted some gaps in current legislation' (UK Parliament, 2021: 16). This line implicitly refers to the tactics adopted by the environmental movement XR, which has organized peaceful, joyful yet highly disruptive protests all around the UK (and also internationally) since 2018. Indeed, their protests have involved the occupying of roads outside Parliament and the blocking of bridges, key intersections and railway services in Central London, with people also gluing themselves to, or locking themselves onto, various objects. In April 2019, for example, thousands of people occupied key sites in London for two consecutive weeks, including Piccadilly Circus, Oxford Circus, Marble Arch, Waterloo Bridge and the area around Parliament Square, causing widespread disruption in the capital (Taylor and Gayle, 2019; see also Gov.uk, 2022).

The reference to XR is made explicit in a report – on which the bill is based – published on 11 March 2021 by Her Majesty's Inspectorate of Constabulary and Fire & Rescue Service (HMICFRS) that investigated how effectively the police manage protests. The report was commissioned by the home secretary on 21 September 2020 '[f]ollowing various incidents that had caused disruption – including protests by Extinction Rebellion and Black Lives Matter' (HMICFRS, 2021: 17). The report specifically mentions the protests organized by Black Lives Matter, XR and other environmental groups that have organized protests against fracking, the badger cull and the construction of the high-speed railway line HS2. In relation to XR in particular, the report acknowledges that '[i]n April and October 2019, Extinction Rebellion brought some of London's busiest areas to a standstill for several days', blocking the delivery of newspapers and actively seeking 'arrest, in an attempt to overwhelm the police and justice system' (HMICFRS, 2021: 17). The report acknowledges that all these protests have been peaceful, yet it frames them as instances of 'low-level aggravated activism', which 'involves unlawful behaviour or criminality, has a negative impact upon community tensions, *or causes an adverse economic impact to businesses*' (HMICFRS, 2021: 21, emphasis added). In essence, the report and the bill try to strike a new balance between the right to protest and the general interests of the community, with such an effort arguably being needed to counterbalance the disruptive effects that the former has increasingly caused for the latter and to businesses in particular. In the HMICFRS report, this new balance tilts in favour of new police powers to control protest, which the report argues should also include intrusive surveillance mechanisms, such as live facial recognition and covert intelligence.

The part of the bill related to public protest was subject to fierce criticism by various civil liberty groups (Siddique, 2021), which succeeded in initially postponing the parliamentary discussion (Shennan, 2021). Even the House of Lords asked the government to rethink some of the provisions related to protest and rejected the most contentious ones, which were mostly introduced through amendments during the various stages of the parliamentary discussion.[10] Currently, the UK government is considering a further increase in police powers to deal with public protest, specifically to counter the locking-on, gluing and other protest tactics used by environmental groups such as XR and Just Stop Oil (Gayle, 2022a; Townsend, 2022; Mason et al, 2023).

Similar to England and Wales, many European countries and cities have also increasingly come to view protesting as an activity that often causes disruptions to businesses and economic activities. Protesting, to be sure, is often disruptive, and the recent initiatives by, among others, XR, Insulate Britain and Just Stop Oil in the UK prove this point. However, the timing matters: it is no coincidence that the right to protest in many countries has come under attack specifically during the pandemic – a time of deep economic uncertainty and strain. Indeed, during the COVID-19 pandemic, peaceful protesting has been limited not only for public health concerns but also often to safeguard economic interests in consumption-focused inner-city areas. An example is provided by the city of Vienna, which, in December 2021, considered banning daytime protests in the city centre for the following reason: there had been 'several weekends of demonstrations … against Covid-19 measures', and there was therefore the need 'to ensure businesses are not interrupted during the last weekend of trade before Christmas, with Vienna city centre expected to be busy with shoppers and motorists' (Maguire, 2021). Economic reasons also motivated the bans on the so-called 'Freedom Convoys' in European cities in February 2022 (BBC, 2022; Rauhala and Aries, 2022), which opposed COVID-19-related mandates and restrictions. These bans were motivated by the need to prevent the serious disruptions that had just happened in the Canadian city of Ottawa, where 'Freedom Convoys' blocked the city centre through vehicles and trucks, causing sustained and substantial disruptions to traffic, trade and the general economy (see, for example, France24, 2022).

Similar to Vienna, in Italian cities, protesting in city centres has also often been viewed as a hindrance to economic and commercial interests. As previously discussed, my research in Trento demonstrated how eco-justice protests tended to be allowed by the police only outside the city centre, while other demonstrations (probably considered less problematic by the police) could be held in this area at the same time. This gradually changed towards the end of my fieldwork: early in November 2021, the local *questura* decided to also displace 'disruptive' anti-COVID-pass protests outside the centre, and shortly after, it was also followed by the Italian Minister of the Interior Luciana Lamorgese, who allowed prefects to identify 'sensible urban areas, of particular interest to community life', where all protesting could be prohibited (Interno, 2021). As the minister explained in her directive of 10 November 2021 (Interno, 2021), the need to more tightly regulate protest came from the anti-COVID-pass protests that had been organized across the country for several months. As noted by the minister, frequent anti-COVID-pass protests not only posed a public health risk but also had 'negative effects particularly on the current phase of gradual recovery of the social and economic activities' in the country (Interno, 2021). The economic rationale of this directive was confirmed by news reports, which emphasized the loss of revenues by businesses and retailers in city centres due to the frequent and highly crowded anti-COVID-pass protests (*TheLocal.it*, 2021; Ziniti, 2021). The directive did not force prefects to identify 'no-protest' zones within their territory of competence, but it possibly gave them a push in that direction, while also emboldening those who had this practice already in place before the pandemic began (such as the *questura* of Trento for eco-justice protesting, as discussed earlier). As Ziniti (2021) anticipated, after the directive, protests in Italian cities would most likely be allowed only when held outside city centres, away from shopping streets and 'sensitive' sites (such as institutional or political headquarters), during times when they are least disruptive, and only in 'static' form (that is, not involving parades, processions or marches).

Drawing on recent examples mostly taken from the European region during the COVID-19 pandemic, this section has illustrated how the need to protect economic and commercial interests, particularly in inner-city areas, negatively affected the right to protest. Protesting has come under attack for two main reasons: either because protest strategies have allegedly become seriously disruptive to commerce and travel (as in the case of XR in the UK or of 'Freedom Convoys' in Canadian and European cities); or because protesting has come to be viewed as a hindrance to the local economy and the post-lockdown economic recovery (as in the case of the Italian new ministerial directive). In both cases, it is apparent that the motives underpinning policy changes around protesting are fundamentally economic: with such policies, national and local governments strive to protect commercially valued areas from behaviour than can potentially disrupt businesses, discourage consumption and therefore undermine local economic revenues. As a result of that, inner-city areas are increasingly becoming less welcoming of dissent, critical views and political struggles.

Limiting protest in public spaces: some considerations

Drawing on multidisciplinary and interdisciplinary insights, this chapter has illustrated how the ethos of consumption imbued in inner-city spaces has fundamentally shaped individual behaviour and led to the exclusion of unwanted bodies and embodied practices from sight. Indeed, city centres in Europe have increasingly come to be dominated by 'order', normativity and exclusivity – traits that have also recently affected the right to protest. Increasingly understood as causing serious disruptions and an impairment to the post-lockdown economic recovery, protesting has been pushed out of city centres (as in the case of Italy) or been more tightly controlled and punitively sanctioned (as in the case of the recent Police, Crime, Sentencing and Courts Act 2022 in England and Wales).

With little to no protesting in city centres, however, these public spaces ultimately relinquish their function as a public sphere – connecting people and allowing them to meaningfully engage with each other, including by challenging the status quo when needs be. In other words, without protests, city centres tend to become rather sterile and unimaginative: spaces where the spontaneous production of alternative imaginaries around the meaning of the city and citizenship is stifled, and where the role of public spaces as sites of consumption is ultimately emphasized (Arenas and Sweet, 2019b). These spaces should not be uncritically accepted but, rather, resisted – as they often are – by individuals and social movements in an effort to (re)create spaces of and for political expression in urban centres. As Smith and Low (2006: 16) eloquently put it in the introduction to their edited collection *The Politics of Public Space*: 'whatever the deadening weight of heightened repression and control over public space, spontaneous and organized political resistance always carried with it the capability of remaking and retaking public space and the public sphere'. It is precisely to spontaneous and organized eco-justice resistance, and the potentiality of its associated imaginaries, that this book now turns.

THREE

Atmospheres of Eco-Justice Resistance During the Pandemic

Introduction

This chapter moves away from the dynamics of power described so far and specifically focuses on activists' practices of resistance against power. It is as equally interested in activists' visual practices of resistance (exemplified by the writing of graffiti and tags, or the use of stickers, stencils, flags and flyers around the city) as it is in their performative practices (which include rallies and flash mobs, among other things). In particular, the chapter conceptualizes eco-justice visual and performative practices in the city as crucial 'affective conveyors' of knowledge around eco-justice harms and injustices. Captured through our senses, activist practices of resistance are, indeed, affectively registered by us before being cognitively processed and responded to. Obviously, activist practices do not equally affect everyone; actually, they may not affect some people at all. For example, they may not affect those who drive through the city sitting in their SUVs, those who walk compulsively checking their phones or generally those who navigate the city without sensing and being attuned to their surroundings. By the same token, green visual resistance and street-based protest may affectively 'move' some other people, leading to more complex and alternative readings of public spaces and their function, everyday meanings and imagined futures. In short, by mobilizing the senses and affect, urban visual and performative resistance may help generate a different 'sense of place' – one that champions eco-justice and that is ultimately more respectful of the environment and non-human species inhabiting the planet alongside us.

Focusing on the senses and affect in urban public space seems relevant and topical particularly in pandemic times, when (at least initially) human presence in urban space was called into question (with messages saying, 'Stay at home!'), our behaviour was more tightly regulated and our sensorial experiences were seriously curtailed (Young, 2021). Even during this time, however, the sensorial and affective dimensions were never abandoned or proved useless in making sense of our experiences of and in the city; rather, as Young (2021) suggested, they helped us capture newly formed 'atmospheres of control' in the city, as well as their subversion and resistance.

In recent years, criminological scholarship has paid increasing attention to the senses, affect and atmospheres when addressing crime, crime control and justice (see, for example, Herrity et al, 2021; Peršak and Di Ronco, 2021). As a 'spatialised feeling' that 'connect[s] people and place in a shared experience' (Fraser and Matthews, 2021: 456), atmospheres have been recognized as a useful concept enabling a multilayered understanding of control and power relationships in criminal justice settings (Young, 2019), as well as in the city or street (Young, 2021; Fraser, 2021; Fraser and Matthews, 2021). Atmospheres are, however, never fixed; rather, as the scholarship on atmospheres instructs us (see, for example, Anderson, 2009; Sumartojo and Pink, 2018), atmospheres are emergent, evanescent, dynamic, flexible, fundamentally ambiguous in their essence and hence also open to sudden ruptures and challenges by other emergent atmospheres. Due to their innate ambiguity, atmospheres have the capacity to introduce contestation into established atmospheres of order, control and discipline, while also suggesting alternatives to them (Young, 2019, 2021; Fraser and Matthews, 2021).

This chapter contributes to 'critical sensory criminology' (McClanahan and South, 2020) and the criminological scholarship on atmospheres by focusing on the atmospheres of green visual and performative resistance during the pandemic. It draws on the flexible approach developed by Sumartojo and Pink (2018), who posited that atmospheres always-already exist and hence that we make sense of our social world while being immersed *in* atmospheres. As these authors cogently put it, once in and attuned to atmospheres, the challenge is to find the best terms to describe them (or 'know about' atmospheres) and understand which previously un-envisaged scenarios they make possible for us (or 'know through' atmospheres).

To 'know about' and 'through' atmospheres, in this chapter, I engage in three autoethnographic exercises, where I approach some of the affective atmospheres to which I became attuned during fieldwork. As mentioned earlier in this book, I conducted fieldwork in the city of Trento from November 2020 until August 2021. For my three exercises, I used my fieldnotes and the photos I took during fieldwork to reflect on the atmospheric configurations that affected how a place (a bridge) and an event (a street-based protest) felt and meant to me, while also producing a sense of intimacy that invites emphatic imaginaries of what it was like to be there (Pink, 2015). Ultimately, these efforts serve the purpose of demonstrating how being in atmospheres of resistance enables knowledge of eco-justice harms and imagination of future transformative scenarios that foreground the health and well-being of non-human animals and the environment, and prioritize them over unsustainable and harmful human economic development, growth and progress.

This chapter starts by reviewing the criminological literature attuned to the sensory and affective dimensions of deviance, crime, crime control and justice. It then focuses on my three autoethnographic exercises after having outlined their methodological background. Drawing on these exercises, the chapter suggests future research avenues for future green critical sensory criminological research on eco-justice resistance and its atmospheres. As I will argue, such research is important as it can reveal past, ongoing and future eco-justice harms and open up possibilities for future progressive and sustainable living that are more attuned to the more-than-human surrounding us.

The senses, affect and atmospheres in criminology

Over the last decade, criminology has increasingly recognized the importance of the senses, affect and emotions in the study of crime, harm, crime control, punishment and justice. Among the many existing criminological perspectives, cultural criminology has provided the most fertile ground for this new endeavour to thrive, with the pioneering work of Jack Katz (1988) focusing on the emotional foreground of criminality. It is within this approach that, for example, authors have suggested innovative ways of looking at space and crime (Hayward, 2012), and have launched the burgeoning field of visual criminology (Brown and Carrabine, 2017), which later inspired efforts to also include other senses beyond sight in criminological research. Empirical and theoretical developments in these areas have contributed to giving new impetus to criminological research and inspired new (critical and cultural) criminological endeavours more attuned to the sensory, affective and emotional dimensions of crime, crime control and justice (for some recent examples, see, for example, Herrity et al, 2021; Peršak and Di Ronco, 2021). This section reviews the recent developments in criminology concerning the intersection between deviance, crime and crime control, the senses, and affect, starting with the studies that offered new conceptual lenses through which to approach space and spatiality within cultural criminology.

In a key agenda-setting article, the cultural criminologist Keith Hayward (2012) called on criminologists to take space seriously – that is, to think about space no longer as a mere geographical background to crime but as a phenomenological locus of power relations, cultural and social dynamics, and everyday values and meanings. To help criminologists think more creatively about the relationship between space and crime, in his article, Hayward (2012) outlined 'five spaces'. Of particular importance to this chapter are the first and last types of spaces he pinpointed. The first are the so-called 'more-than-representational spaces', which are drawn from non-representational theory (NRT) within cultural geography. As put by Hayward (2012: 449, emphasis added), NRT is 'an attempt to move beyond static geographic accounts of landscape in a bid to create an alternative approach that actively incorporates the *experiential, affective, and intermaterial aspects of space*'. In essence, Hayward (2012: 451, emphasis in original) alerted cultural criminologists to the importance of analysing 'affective landscapes or experiential places' where 'emotions actually influence or shape space more than they arise *in* space'. Moreover, in his fifth and last suggested spaces, called 'soundscapes and acoustic spaces', Hayward emphasized the importance of the aural dimension – and of the sensory dimension of crime and control more generally. The latter spaces have later been drawn upon by McClanahan and South (2020) in their efforts to discuss a 'critical sensory criminology', to which I shall return later.

Responding to Hayward's (2004, 2012) work on the crime–city nexus, Campbell (2013: 23) focused on urban 'crime' and understood it as an affective and performative event that stimulates alternative or renegotiated senses of place – as something that challenges the meanings and uses of space, which are 'always-already in flux'. As Campbell (2013: 35) put it: '"Crime" has affective power which excites, threatens, angers, shocks; it provokes an embodied experience of place and sensibilizes us to change and alterity in everyday settings.' Other cultural and critical criminology scholars have used (more or less explicitly) the idea of 'crime' or 'deviance' as affective mediums that help challenge dominant social orders. Drawing on visual criminology, Young (2014), for example, focused on the affective encounter between the spectator and unauthorized images in the urban space (in her 2014 article, she focused on the painting of an advertisement located in the inner panels of a bus shelter). Young clarifies that affect is something that precedes emotional states and that is key to the process of meaning making: 'affect marks the moment at which connection to something seen, heard, experienced or thought registers in the body and then demands that it be named or defined' (Young, 2014: 162). According to this author, when spotted, images are affectively encountered by the spectator; it is through such affective encounters (and affective relations between the spectator and the image) that meaning emerges and is produced. As Young (2014: 170) argued: 'the image is encountered as something that has changed the space and thus demands the spectator's engagement in its interpretation. Thus, before any particular intellectual or emotional response is engendered, the image seeks the spectator's engagement as an affective intensity.' Indeed, intellectual efforts to make sense of the image, as well as the eventual emotional responses to it, only come after the image is affectively registered by the observer. Young (2019, 2021) developed these ideas further in some of her recent work, where she used the concept of 'affective atmospheres', which she defined as 'that which connects individuals within and to the spaces they occupy and move through' (Young, 2019: 766). Permeating spaces of criminal justice (Young, 2019) or the lockdown city (Young, 2021), atmospheres emanate an affective charge that is registered by subjects through their senses and that helps inform, reaffirm or simply illuminate disciplinary relations between individuals and the state. However, that is only a part of the story. As Young (2019) cogently pointed out, dominant atmospheres are also open to contestation, for example, when they are ruptured by an unexpected element or occurrence that can lead to its renewal or replacement.

Important in this context is also the work of Andrew Millie, who, in recent articles, has suggested the idea of an 'aesthetic criminology' (Millie, 2017, 2019, 2022). Such a criminology draws on both visual and cultural criminology, and 'includes considerations of the visual, but also broader sensory, affective and emotive experience' (Millie, 2017: 4). On his account, urban interventionism (Millie, 2017), including yarn bombing (Millie, 2019) and guerrilla gardening (Millie, 2022), are performative practices that challenge dominant aesthetic orders through mobilizing the senses and, through them, affect and emotions. Similar to Young (2014), for Millie too, intellectual or emotional responses to 'deviance' or, more accurately, taste (in the cultural sense) only come after their sensory and affective reception by the public.

The importance of the senses in criminological scholarship has recently been stressed by McClanahan and South (2020), who suggested the need for a 'sensory criminology' within critical criminology – that is, a criminology concerned with all senses (not only sight) and attuned to the many ways in which these intersect with crime, harm, crime control and justice (for earlier accounts on criminology and the senses, see also Shalhoub-Kevorkian, 2017; Garcia Ruiz and South, 2019). As McClanahan and South highlighted, moreover, the senses are necessarily also conveyors of affect and affective atmospheres; as they argue: '"atmospheres" are already always spaces configured in the totality of sensory information, and so cannot be easily reduced to "the visual", "the tactile" and so forth' (McClanahan and South, 2020: 13). The linkage between the senses and atmospheres has also been highlighted by Lundberg (2022) in her discussion of 'phenomenological atmospheres', which she views as entanglements through which ordinary green harms at the expanding urban edges are captured by humans through their senses. Her understanding of atmospheres, however, goes well beyond the phenomenological aspect; in addition to atmospheres as perceived by human senses, she also introduces 'ontological atmospheres' – a category that goes beyond human perception and captures ordinary harms as impacting on the more-than-human.

Important criminological work on atmospheres has also been done by Fraser and Matthews (2021), who focused on the atmospheres of street-based protest. In particular, they focused on the atmospheres generated by Hong Kong's Umbrella Movement, a pro-democracy campaign that occupied Hong Kong's central streets for 79 days in 2014. Using atmospheric methods (Anderson and Ash, 2015), which are non-representational, sensory and subjective, these two authors approached the street (and protest happening there) as both a physical and an affective place – a locus where distinct spatialized affective spheres can be 'forged through a "combinatorial force field" of human and non-human elements (Amin and Thrift, 2017: 16), operating at a level of experience that is affective and infra-conscious, with the capacity to act on individuals in a quasi-agentic manner (Schuilenburg and Peeters, 2018)' (Fraser and Matthews, 2021: 456). For Fraser and Matthews, non-human elements are municipal regulations and the law more in general, technology, architecture, and other material objects, such as protest posters, flowers and plants. Together with human bodies, non-human elements generate a distinct atmosphere during protests through the creation of moods, feelings and sensibilities – a shared affective intensity that has the capacity to disrupt prevailing normative orders.

Such an affective charge is, however, only captured at the edge of an atmosphere, where it dissolves and the prevailing one (re-)emerges. As Fraser and Matthews (2021: 463) put it: 'Indeed, it is precisely at the edge of the encampment's atmosphere that the force of the prevailing atmosphere of docile consumption, which typifies the "legislated" spaces of Hong Kong, was starkly foregrounded, suddenly rendering visible the taken-for-granted backdrop that shapes the everyday normative ordering of the city's streets.'

Fraser and Matthews also acknowledged that protests' affective charge can remain well beyond demonstrations have ended and be reanimated in other related protests or events. Indeed, as the research by Kindynis (2019) demonstrated, atmospheres' affective charge can be reminisced long after it was generated: it can haunt abandoned, isolated or forgotten places, and emanate from the graffiti, tags and throw-ups written by artists over time and left behind in off-limit spaces. As he contends, these 'lingering "spectral traces" – the absent presences – of those "who were apparently" here "earlier but who have now moved on or are missing"' (Kindynis, 2019: 37) can be foregrounded through an active engagement with the atmospheric resonances of urban space.

Criminological studies on atmospheres have mostly been informed by the work of the cultural-political geographer Ben Anderson (see, for example, Kindynis, 2019; Young, 2019, 2021; Fraser and Matthews, 2021), who defined atmospheres as 'singular affective qualities that emanate from but exceed the assembling of bodies' (Anderson, 2009: 80). According to Anderson, one key characteristic of affective atmospheres is their ambiguity: they are 'collective affects that are simultaneously indetermined and determined', and they are 'never finished, static or at rest' (Anderson, 2009: 78, 79). As Fraser (2021: 221) also concedes, atmospheres are 'produced by a moment that reaches out beyond subjects and "touches" others, connecting self, other and space in a shared experience'; such sharing, however, is 'a fractured, temporary and evanescent alliance of bodies in space, produced by and producing mobile and shifting subjectivities'. As a consequence, atmospheres are not easy to pinpoint, making it a challenge to describe, research and analyse them (Sumartojo and Pink, 2018).

Methodology

Affective atmospheres are, indeed, difficult to research; yet, recent literature, mostly produced within the discipline of anthropology, provides interesting insights that assist the qualitative researcher in this effort, including by encouraging them to take their affects and emotions seriously while doing fieldwork and analysing their collected data (see, for example, Pink, 2015; Kahl, 2019; Stodulka et al, 2019). Particularly important for this chapter is the flexible framework outlined by Sumartojo and Pink (2018) in their 2018 book *Atmospheres and the Experiential World: Theory and Methods*, which is an empirically grounded approach to researching, understanding and theorizing about atmospheres. Sumartojo and Pink (2018: 15) defined atmospheres as the 'ongoing sensory and affective engagement with our lives and their impressions, sensations and feelings and the environments through and as part of which they play out'. In other words, atmospheres for them always-already exist: they are part of the environments we inhabit. This implies that we live and make meaning about our world *in* atmospheres, while also being influenced by our foreknowledge, memories and imagination or anticipation of the future.

Living and being in atmospheres, however, does not mean that we are able to properly grasp their emergence and configurations while we are *in* them. As Sumartojo and Pink (2018: 41) contended, we often 'know about' atmospheres retrospectively and, more precisely, through methodologies that allow us to 'transform ... accounts of the experience of being in atmosphere ... into something descriptive or representational of those atmospheres' (such accounts are however not fixed; rather, they need to account for the ephemerality, dynamicity and multiplicity of atmospheres). As the authors clarify, the stage of 'knowing about' atmospheres is always reflexive and hence inherently autoethnographic: it involves engagement with own experience and hence with the 'atmospheric configurations where we have realised or known something that has enabled our research' (Sumartojo and Pink, 2018: 39).

'Knowing about' atmospheres also leads to 'knowing through' atmospheres, or grasping how the atmospheres we live in and may know something about may have an impact in our lives, for example, by actively shaping our understanding of the world and its futures. As Sumartojo and Pink (2018: 45) argued: 'atmospheres also can ripple forward in time, carrying the terms of experiencing them into the future, creating a new set of possibilities that were not knowable or possible in the same way before orienting towards atmosphere'.

Drawing on Sumartojo and Pink (2018), in this chapter, I engage in three autoethnographic exercises in an attempt to 'know about' and 'through' specific atmospheric configurations after having being in them. In particular, through these exercises, I endeavour to reflect on the qualities that affected how a specific event (a street-base protest) and space (a bridge) felt and meant to me while doing ethnographic research, while also attending to the future imaginaries they suggested.

The first two autoethnographic exercises are focused on a street-based protest organized by the NOTAV Committee on 15 May 2021. This was the first in-person protest I attended during fieldwork (previously, I had attended various online meetings of the committee and its assemblies), so I had no specific expectation of how the protest would unfold or be organized; however, I did have foreknowledge on protest policing practices from my previous research and conversations with activists.

The space I considered for the third autoethnographic exercise is the San Lorenzo bridge – a bridge in Trento crossing the Adige River. During fieldwork, I regularly walked over this bridge, as well as for many months afterwards (my home is located just after the bridge in the Piedicastello neighbourhood). The bridge connects the city centre and its main train station to the Piedicastello neighbourhood, which is also the area where the local social centre is located.

In the following first two sections, titled 'Atmospheres of resistance' and 'Ruptured atmosphere', respectively, I use the example of a street-based protest to explore the concept of 'affective atmosphere' and its sudden rupture. Following Fraser and Matthews (2021), I used atmospheric methods (Anderson and Ash, 2015) and, concretely, fieldnotes as prompts to discuss the affective charge of 'entanglements' between human and non-human bodies during the protest, along with their sudden rupture and the imagined scenarios they opened up for me. The resulting texts – fieldnotes followed by commentaries – produced an 'ethnographic place' (Pink, 2015): a place where I tried to evoke the corporeal and experiential feeling of being in the atmospheres coalescing around the protest, while also hopefully lending to the reader an empathetic understanding of what it was like to be there.

In the third section in the following, titled 'Atmospheres of visual resistance', I used the fieldnotes and pictures[1] I took during fieldwork as prompts to evoke my memories and imagination around the affective encounters I had with markers of green visual resistance while walking on the San Lorenzo bridge. The use of fieldnotes and pictures as prompts enabled me to conjure the embodied and emplaced knowledge that the research situation involved for me (Pink, 2015). In other words, in order to 'know' something 'about' and 'through' such encounters, I paid specific attention to my sensory memory (Pink, 2015) – that is, my memory of the senses that triggered an affective relation with visual resistance in space and that unbridled my imagination around future change. The resulting reflexive and evocative text, and the commentary that followed it, are also in themselves an 'ethnographic place' (Pink, 2015): a place where written description, visual evocation and theory are brought together and interwoven, while simultaneously inviting the reader to an empathetic understanding of my own emplacement (that is, the sensory and affective relationship that I experienced between the body, the mind and the environment during fieldwork).

In all three exercises, walking was central to my sensing and experiencing of (and attuning to) atmospheres, allowing me to capture their intensities, qualities and future potentials. Movement is recognized as central in much of the existent literature on atmospheres, which sees it as an embodied change that is crucial to the constituting, making and experiencing of atmospheres (see, for example, Sumartojo and Pink, 2018; Riedel, 2019). Walking is also valued in recent innovative cultural and critical criminological research, where it is used as a multi-sensory, creative or 'imaginative' (Seal and O'Neill, 2021) method to collect people's stories and lived experiences of transgression and harm in space and place (see, for example, Natali, 2019; Natali and de Nardin Budó, 2019; Neville and Sanders-McDonagh, 2019; O'Neill and Roberts, 2019; Natali et al, 2021; Seal and O'Neill, 2021).

FIELDNOTES
Atmospheres of resistance

As I cross the street and I walk into the San Martino neighbourhood, I feel like I am entering another world: I leave behind the very polished historical centre, buzzing with the Saturday-afternoon shoppers, and I enter an equally good-looking street with bars and nice shops but being 'made up' as if it was a construction site. Already from the other side of the street, I could see a red safety net over the metal pedestrian barrier and three stilt walkers and other people wearing helmets and high-visibility clothing (yellow-and-orange vests).

While I walk into the street, stilt walkers dressed like builders are moving around in an effort to mark the edge of the street with yellow-and-black tape – the same tape used to encircle construction sites (see Image 3.1). As I step foot into the neighbourhood, I observe stilt walkers encircling the edge of the San Martino street (the tape is being pulled just above my head) and tying the tape around road signs, flag poles, a building's gutter, a bar's outside umbrellas and lamp posts (see Image 3.2). Around me, 20 other people, of all ages, are also observing these activities. A family with a boy is standing next to me; he looks at the scene quite thrilled. The mood is very cheerful and joyful, especially because of the presence of the stilt walkers moving around and taping and encircling the area.

The edge of the street, where I am standing, is also filled with NOTAV flags, a banner saying 'San Martino from neighbourhood to construction site: No thanks' (see Image 3.3), and the typical signage of construction sites (including a ban on walking around the TAV construction site, a sign with the caption 'excavation danger', and so on). On my right, just in front of the first bar on the San Martino street, there is a desk with flyers and booklets from the NOTAV Committee, where activists in yellow-and-orange vests provide information on the TAV project. I briefly speak with one of the organizers at the desk before the event starts.

The event begins when the sound is switched on: we hear noise typical of construction sites, which helps us feel and imagine how the area will become when (and if) the TAV project will be approved and the construction of the underground tunnel will start. Such noise is alternated

POLICING ENVIRONMENTAL PROTEST

Image 3.1: Stilt walker pulling yellow-and-black tape across road signs to encircle the protest area

Image 3.2: Tape tied across a bar's outside umbrellas

Source: Anna Di Ronco

Source: Anna Di Ronco

Image 3.3: Banner at the entrance of the San Martino neighbourhood

Note: banner reads 'San Martino from neighbourhood to construction site: No thanks'
Source: Anna Di Ronco

> by Italian songs championing eco-justice and supporting resistance against oppression, and by the recorded voice of activists informing the public about the TAV project and its negative impacts on the area. In particular, the recorded voices mention air and noise pollution and increased car traffic caused by the construction sites; they also speak of structural damages to buildings linked to the underground tunnel construction and of the further contamination of groundwater by dangerous chemicals already present in the soil along the project's route.
>
> As the recorded tape ends, we start walking towards two other areas of the neighbourhood that will be affected by the TAV project. As we do so, stilt walkers pull the tape from the neighbourhood's edge (where the event started) and attach it to street signs, buildings' gutters and so on up until the next stop.

This fragment is taken from my fieldnotes at the very beginning of the NOTAV event on 15 May 2021.[2] Like many eco-justice protests, flash mobs and other playful events and interventions (see, for example, Millie, 2017; Brisman and South, 2013, 2014), this event had a playful, joyful and cheerful vibe, which was given to it mostly by the carnivalesque presence of stilt walkers. However, there was something more to it – something that I was not quite able to pinpoint at first but that I could grasp through the senses and register through affect while being at the protest. This something was also addressed by Anderson (2009) while discussing the term 'atmosphere'; as he argued:

> [p]erplexingly the term atmosphere seems to express something vague. Something, an ill-defined indefinite something, that exceeds rational explanation and clear figuration. Something that hesitates at the edge of the unsayable. Yet, at one and the same time, the affective qualities that are given to this *something* by those who feel it are remarkable for their singularity. (Anderson, 2009: 78, emphasis in original)

Indeed, although I found it difficult at the time to exactly pinpoint and articulate the affective qualities emanating from 'envelopments' (Anderson, 2009) of human and non-human bodies, I was able to perceive them through the senses and register them through affect. In particular, the senses were essential to mobilize affect and enable knowing in and about atmospheres: seeing activists in yellow-and-orange vests, yellow-and-black tape being pulled around and the signage of construction sites, as well as hearing the noise of construction sites, conjured a sense of knowing what it would be like to live in the neighbourhood once the works for the TAV start. In other words, being in such an atmosphere enabled me to imagine the transformations that will likely occur in the neighbourhood should the TAV project be initiated, as well as the harms it will likely cause. At the protest, therefore, I could gain knowledge of TAV-related environmental harms not only through overt and direct communication by activists, but also at the infra-conscious (sensory and affective) level through the staged yet embodied experience of being in a construction site – an experience that enabled me to imagine the gloomy future of the neighbourhood and that emboldened my opposition against the megaproject.

FIELDNOTES
Ruptured atmosphere

When we get to our second stop, Largo Sauro, I notice one local police car and two local police officers watching us from the edge of the square. I had already noticed them at the beginning of the event, standing just around the corner, in a square 15 metres away from the San Martino street. Here, in Largo Sauro, I also notice a few plainclothes DIGOS agents looking at the crowd. I register them as DIGOS agents as they are standing at the end of the square looking at the crowd, sometimes taking pictures of us. I know that I am not the only one who notices them at this point. Next to me I hear a young woman telling a friend how sick she is of seeing DIGOS agents everywhere.

As we walk towards our third stop, I talk to Diego (a pseudonym), a NOTAV activist. While we walk, we discuss police surveillance and control of activists, which he says have increased over time. As we speak, a police car and a riot-control vehicle drive next to us. I wonder where that vehicle is driving to. Diego sees me looking at the vehicle and tells me: "They are here for us – they are driving towards our next stop." Diego is indeed right: we will see that vehicle at our next stops, together with the local police car and another car of the state police. (From the social media posts of the NOTAV Committee, published after the event, I also learnt that the riot-control vehicle had been with us the entire time; it was more hidden at the very beginning of the event and became more visible towards its end, when I noticed it.)[3] It is at that point that someone tells Diego that DIGOS agents require us to wrap up our flags while waking. This is needed to be able to comply with the national COVID-19-related regulation that only allows static protests to take place. Apparently, the head DIGOS agent acknowledged the nonsensical nature of that requirement but had asked for compliance with it nevertheless. People around me looked quite displeased while wrapping up their flags, but I did not hear them making many comments.

Our last two stops are at Cortile Le Fornaci and Scalo Filzi. It is only here that I fully appreciate the scale and nature of police surveillance around us – watching us. There are two local police agents with one car, one riot-control vehicle, one car of the state police and at least seven plainclothes DIGOS agents. I notice some DIGOS agents to the left of the crowd and some others to the right. If they were not so noticeable in the earlier stages of the protest to a trained eye, they were now visible to everyone. DIGOS agents are standing in groups of three to four people and take pictures and videos of all of us. They also take videos and pictures of three activists while they cross the busy road and hang a large banner on the fence of Scalo Filzi. The banner reads: 'Recovery: mega-projects and ecologic lie' (see Image 3.4).

At this point, the atmosphere gets tenser, despite the cheerful presence of the stilt walkers (they are still with us) and the reassuring one of children and old people. We are looking at the three activists while they hang the banner and know the police are watching us. The tension is almost palpable. I wonder what DIGOS agents will do with all that visual material they are collecting. I also wonder what would trigger an intervention from them; after all, hanging a banner on private or public property can be considered vandalism. We stay there for a few minutes, perfectly aware that the police are watching us closely, taking pictures and videos of us. We then walk back together, with our flags wrapped up. While walking back, I observe some activists removing the flyers, banners and tape from the area of the protest to comply with the set police requirements.

These fragments of my fieldnotes indicate three moments when the atmosphere described in the previous section was ruptured. In all three instances, it was the sight of the police that, to various

Image 3.4: Large banner hung by activists on the fence of Scalo Filzi

Note: banner reads 'Recovery: mega-projects and ecologic lie'
Source: Anna Di Ronco

degrees, defused the atmosphere's intensity and capacity to affect (Anderson and Ash, 2015). In the first two moments captured by my fieldnotes (reporting on the second and third stops of the protest), the sight of the police reduced the atmosphere's capacity to affect without seriously undermining it: I noticed an increase in police presence, but this was not enough to lessen the affective charge generated by the entanglements between bodies and objects constituting the protest's atmosphere. This changed at the end of the event, when the atmosphere's capacity to affect was seriously weakened by the more visible presence of police vehicles and plainclothes DIGOS agents taking pictures and videos of the crowd.

According to Fraser and Matthews (2021) and Young (2019), atmospheres tend to be imperceptible but become apparent at their edges, for example, when entering or exiting them, or when they are ruptured by an unexpected and sudden occurrence (see also Anderson and Ash, 2015). Indeed, in this case, the protest's atmosphere also only became evident – almost palpable – in two instances: when walking in it for the first time (see the previous autoethnographic exercise); and when it was ruptured by the sudden materialization of police forces around protesters. While in it, the atmosphere felt like a 'hermetically sealed bubble' (Fraser and Matthews, 2021: 463): only what was 'inside' the protest seemed to matter and everything 'outside' was ultimately forgotten. That changed suddenly: the joyful and carnivalesque atmosphere that sensorily and affectively foregrounded TAV-related environmental harms was ruptured by the sudden sight (and sudden realization of the extent) of police surveillance and control at the event. These ruptures defused the atmosphere's capacity to affect – and they did so particularly towards the end of the event. Indeed, during the last stop and for a few moments, the protest's affective charge was overridden by another atmosphere and its affective qualities. The resulting dominant '"bubble ..." of shared feeling' (Fraser, 2021: 221) was one that foregrounded the fear of police surveillance and intervention, and possible reprisals after the event. The fading of the protest's atmosphere coincided with the emergence of a new one – one dominated by police power and control. Overall, the protest's atmosphere was never completely overridden and thus maintained an affectively discrete quality throughout; yet, it was seriously weakened by this sudden transformation and never quite recovered from it.

FIELDNOTES
Atmospheres of visual resistance

The neighbourhood of Piedicastello can be spotted from the San Lorenzo bridge crossing the Adige River (see Image 3.5). The river is magnificent. Its large riverbed accommodates water flowing at various speeds towards the city of Verona. Its water is crystal clear but becomes muddy and flows at a greater speed after heavy rain. Along the two riverbanks, two stripes of grass and trees follow the shape of the river and stretch both north and south; they also host the pedestrian path and bicycle lane.

On the bridge, pots with colourful flowers are found in every season, spreading their floral scent during the spring and summer. The wind often blows on the bridge gently or, at times, forcefully, blowing the flowers as well as my hair and those of the other pedestrians walking along the bridge.

Just next to the flowers, stickers of all sorts populate lamp posts. Among the stickers that I observed more frequently while waking (and that were often immediately replaced after they were removed or damaged by the weather) were those with 'MISSING' wild bears as captions. With blue, orange or pink backgrounds, these stickers included the image of a wild bear and some captions, including 'MISSING' at the top and 'STOPCASTELLER' at the bottom (see Image 3.6). These stickers told the stories of wild bears who were killed by humans and, for example, were drowned during capture or were sleep-aid poisoned. Seeing these stickers on the bridge – with the flowing river in the background, the sight and smell of flowers in the pots, and the wind blowing through my hair – left an impression on me: it made me sensorily and affectively connect with the non-human world and think specifically of non-human animals that, like wild bears in this instance, are often harmed by humans. It also made me wonder what the city will look like in the future and whether it will still involve disciplinary and punitive attitudes towards wildlife, or rather have a more positive and inclusive relationship with non-human animals populating this world alongside us.

Seen from the bridge, the Piedicastello neighbourhood is often dim: it is overlooked by the rocky yet green hill called Doss Trento, on the right, and by the Bondone Mountain, on the left. The sight of the Doss hill's steep rock dotted with wild plants and shrubs sparks in me an affective reminiscence: it was the place where a large NOTAV flag was hung and laid for 15 consecutive days in June 2021 (see Image 3.7). The flag was hung in occasion of the 'Festival of Economics', a festival that takes place in Trento every year. For the 2021 edition, the then Italian Minister for Ecological Transition Roberto Cingolani was invited to attend. The minister is one of the members of government most involved in the implementation of the PNRR, which funds the TAV project in Trento. The large NOTAV flag was hung on 5 June and was only removed 15 days later (on 20 June) on occasion of an important local festival (Feste Vigiliane). Until it was removed, the flag was very visible in the area. Seen from the bridge lying against a protruding rock – with the senses being stimulated by the river, the wind and the flowers – the flag acted as a potent affective and cognitive reminder of the opposition against environmentally impactful megaprojects like the TAV.

Overall, for me, the bridge is a place that invites alternative and transformative readings of the future of the city. In the scenarios for future living that I imagined while walking on it, humans abandon their destructive plans to forward progress at all costs; they also relinquish their assumed and imposed centrality to foreground non-human well-being and the health of the environment.

Image 3.5: The San Lorenzo bridge

Source: Anna Di Ronco

Image 3.6: 'MISSING' wild bear sticker on a lamp post on San Lorenzo bridge

Note: sticker reads 'MISSING | KJ2G1 | BROWN BEAR, 2,5 YEARS OF AGE, KILLED IN 2008 BY DROWNING DURING CAPTURE | #STOPCASTELLER'
Source: Anna Di Ronco

Image 3.7: NOTAV flag on Doss hill as seen from San Lorenzo bridge

Source: Anna Di Ronco

In this reflexive and evocative exercise, I conjured my memory of, and imagination around, a number of affective encounters I experienced while walking on the San Lorenzo bridge during fieldwork, attending in particular to the sensory cues that triggered them. As Sumartojo and Pink (2018) argued, memory and imagination are both important when approaching atmospheres: they shape how we constitute, experience and make sense of atmospheres while being in and knowing about them, and work towards taking them forward in time in an effort to know through them. Crucially, as Pink (2015) reminds us, memory is sensory and thus embodied and continually reconstituted through practice.

This autoethnographic exercise addresses my sensory memory of two affective encounters I had while walking on the San Lorenzo bridge: one with some stickers focusing on 'MISSING' wild bears; and one with a large NOTAV flag hanging on a hill's steep rock. In both instances, it was not simply their sight that triggered my engagement with them in an 'affective intensity' (Young, 2014); rather, the senses of sight, hearing and smell were all conjured. These senses were stimulated by the natural environment around which the stickers, the flag and myself as the researcher were immersed: the mountain, the hill, the river and the flowers. In particular, the sight of the mountain and the hill, the smell of floral scent, and the sound of the river flowing and of the wind blowing were all captured through the senses and contributed to mobilizing affect in the service of environmental and species justice. After this affective intensity was generated, cognition came into play (Young, 2014): a process of meaning making addressing the stickers and flag, and answering the questions: 'What do they mean?'; 'Who put them there?'; and 'Why are they there?' Given my foreknowledge of the social context, its eco-justice struggles and movements, it did not take me long to acknowledge the political nature of these markers and speculate on their ownership and meaning (however, it is worth clarifying that this meaning-making process may have had a different outcome for another person and different reactions from them too) (see, for example, Young, 2014; Sumartojo and Pink, 2018). In line with previous literature discussing urban crime (Campbell, 2013) and deviance (Millie, 2017, 2019, 2022), the visual practices considered here also challenge dominant social and aesthetic orders through mobilizing the senses and affect; while so doing, they also stimulate an alternative reading and sense of place – one that invites attention to the harms caused to non-human animals and the environment, and that champions environmental and species justice.

Discussion: knowing in, about and through atmospheres of resistance

In this chapter, I have engaged in three autoethnographic exercises where I reflected on some of the affective atmospheres I attuned to during fieldwork, specifically attending to the senses that were triggered while being in them. Using fieldnotes and pictures taken during fieldwork (which, in the third exercise, were employed to evoke my own sensory memory and imagination of affective encounters with green visual resistance on a bridge), I created three distinctive ethnographic places (Pink, 2015) where written text, theory and visual evocation came together, hopefully also lending a sense of embodied knowledge to the reader. Such an effort served the purpose of knowing about and through atmospheres, after having being in them (Sumartojo and Pink, 2018).

As I illustrated earlier, while being in atmospheres, visual and performative resistance triggered the senses (which, in the case of visual resistance, were further stimulated by the surrounding natural environment); through the senses, affect was also mobilized, enabling the reception of messages at the infra-conscious level – messages that, in this specific case, championed environmental and species justice. In particular, for me, atmospheres of resistance emanating from the street-based protest and from visual resistance on the bridge operated as an affective conveyor of harms caused to wild bears and the environment. Being in them and attuned to them also instilled in me an alternative sense of place – one that suggested a more respectful and positive attitude to the environment and non-human animals in the city, and that prioritized the healthy environment and human and non-human well-being over ruthless economic progress. Indeed, knowing through atmospheres offers the possibility of feeling and thinking differently about the world we inhabit, allowing imagined future scenarios to shape the way we feel and also make decisions in the present (Sumartojo and Pink, 2018). This often happens in spite of sudden occurrences that have the power to defuse atmospheres' intensity and capacity to affect, as the example of visible police presence during the NOTAV eco-justice protest demonstrated. In this example, the atmosphere of surveillance and control that emerged during the latter stages of the protest shone a light on the disciplinary relations existing between activists and the state (on this point, see also Young, 2019, 2021; Fraser and Matthews, 2021). Although this atmosphere of control weakened the atmosphere of resistance emanating from the protest, it did not succeed in erasing it, leaving intact the imaginaries and possibilities it hinted at – possibilities that were also suggested by the markers of visual resistance seen on and from the bridge.

The ethnographic places I produced in this chapter are obviously subjective experiential accounts of what specific atmospheres of resistance felt like and meant to me – a product of my reflexive awareness, sensory memory and imagination of being in them. Obviously, I do not wish to assume that such representations are valid for other people, who may have been equally immersed in atmospheres (perhaps the same ones I have written about) but have not been asked to participate or collaborate in the research project. Involving participants would be an interesting step to take in future post-pandemic criminological research, however, as it would contribute to exploring how people who are themselves in atmospheres of resistance (including protests and areas filled with markers of visual resistance) feel and make sense of them, as well as retrospectively, and the conditions that enable them to imagine possible futures not previously envisaged.

I believe criminology to be a good platform from which to accommodate such 'imaginative' research (Seal and O'Neill, 2021) – ideally, in alliance with other disciplines, as I will discuss later. As mentioned in earlier parts of this chapter, there is a rich literature within cultural and critical criminology that has studied the sensory, affective and atmospheric dimensions of 'deviance', 'crime', harms and the spaces where they occur (Young, 2014, 2019, 2021; Millie, 2017, 2019, 2022; Fraser and Matthews, 2021; Herrity et al, 2021; Peršak and Di Ronco, 2021; Lundberg, 2022). Such literature has specifically highlighted the importance of studying these dimensions to provide multilayered accounts of power relations in space, as well as of efforts to disrupt normative and aesthetic orders from below. Working within the concept of atmospheres not only does that but also helps to reimagine new everyday orders in ways that are more prosocial, inclusive, compassionate (see Fraser and Matthews, 2021; Young, 2019, 2021) and, I would argue, attuned to the more-than-human surrounding us (see also McClanahan and South, 2020; Lundberg, 2022). To unlock this transformative potential, future criminological studies ought to expand the 'criminology of atmospheres' (Fraser and Matthews, 2021) not only by focusing on 'atmospheres of control' and their eventual ruptures and contestations (Young, 2021), but also by specifically addressing 'atmospheres of resistance' and the innovative, inclusive and more sustainable possibilities for future living they often unlock.

In addition to conducting such research in collaboration with activists and other members of the public who are (or were) themselves in atmospheres of resistance, future criminological research in this area should also experiment with methodologies that attend to the more-than-human surrounding us. As I demonstrated in one of the autoethnographic exercises

I presented in this chapter, the natural landscape plays a role in (co-)constituting atmospheres of eco-justice resistance. In such an exercise, I indeed showed the importance of, among other things, the river, the wind and the flowers in co-constituting the atmosphere of visual resistance on the San Lorenzo bridge; in addition to seeing markers of green visual resistance, my sensing of the natural elements on and around the bridge helped to mobilize affect in the service of eco-justice, alerting me to the importance of harms caused by humans to non-human animals and the environment. However, attuning to the natural environment while being in the flow of atmospheres of resistance can go further than that; in particular, as Sumartojo and Pink (2018: 40) suggested, atmospheres can be generated during the research process itself, for example, during workshops held in natural settings focused on the harms caused (or likely to be caused) on the more-than-human by human action or inaction. In practice, this means that the more-than-human can intentionally be attended to while being in the atmospheres emanating from workshops or other research activities with participants. Ultimately, being in such (research-facilitated) atmospheres can help participants establish or reinforce a sense of empathy and care for the more-than-human surrounding them, and also reinvigorate existing efforts for the protection of the environment.

Recent interdisciplinary projects provide some interesting examples that can inspire future criminological (and interdisciplinary) endeavours in this area. In the transdisciplinary project Walking with Zenne, for example, a group of academics, artists, locals and interested people attempted to 'be with' the Belgian River Zenne; they did so by walking alongside the Zenne, while, at the same time, being immersed in observing, listening and sharing their sensory experiences with others (Pali et al, 2022). Among the aims of this project was that of documenting the harms to the Zenne – through hearing 'her' own voice – and of activating a process of repair and care for the protection of the river. Although this project did not use the concept nor language of atmospheres, it could be argued that during these walks, atmospheric entanglements were formed between humans and the natural environment that enabled processes of knowing in, about and through atmospheres. Ultimately, this contributed to opening up new imaginaries and possibilities for the future conservation and thriving of the river.[4]

Future criminological research on atmospheres of eco-justice resistance should ideally also involve interdisciplinary collaborations and, within them, envisage collective activities where participants and the research team know in atmospheres – and as a result, about and through them too. As in the preceding example, such activities can be staged along rivers or, more generally, in natural areas affected by harms caused by humans, including in places where harms have not yet materialized but may well do soon, for example, as a result of megaprojects like the TAV discussed in this book. The emergent atmospheric configurations of people, things, natural environment, non-human beings, feelings and imaginaries that are generated from such activities can go a long way to revealing ways of thinking about repair and modes of living that are more respectful of the more-than-human surrounding us. Ultimately, this can also help shape our everyday actions (for example, by reducing our ordinary acts that contribute to environmental destruction) (see Agnew, 2020), mobilize people, and inspire and embolden eco-justice campaigns.

Conclusion

In this chapter, I have shown new, imaginative ways of doing green critical sensory criminological research on eco-justice resistance and its atmospheres. Using three autoethnographic exercises, I have reflected on some of the atmospheres of resistance (and their ruptures) to which I became attuned during fieldwork, specifically attending to the senses that were triggered while being in them, as well as mobilized by the natural environment, which, as discussed, plays a role in (co-)constituting atmospheres. Following the flexible approach by Sumartojo and Pink (2018), such an effort served the purpose of knowing about and through atmospheres, after having being in them. In other words, these exercises helped me to retrospectively reflect on what it was like to be in atmospheres of eco-justice resistance and to gain knowledge of the harms that humans often cause to the more-than-human in the city (that is, to know about atmospheres). However, my experience and knowledge of atmospheres went beyond that: it also helped me to imagine new radical possibilities for the city of the future – a city that is more respectful of the ecosystems around us and that actively endeavours to protect and repair them (that is, to know through atmospheres).

In the discussion, I suggested new ways of doing future critical sensory criminological research in, about and through atmospheres of eco-justice resistance, while also involving activists, other participants and interdisciplinary collaborations. Studying such atmospheres within criminology (and, crucially, within interdisciplinary projects also involving criminologists) is important not only because it allows us to reflect on existing disciplinary relationships and past, ongoing or future harms, but also because it enables us to think of alternatives to the status quo. Such imagined possibilities carry great, radical potential: they can shape our everyday actions, mobilize people and inspire eco-justice campaigns. Ultimately, studying atmospheres of resistance can enable us to embolden activists' radical efforts and imaginaries, and is much needed to counter increasing activist repression in times of ecological destruction and climate change.

Conclusion

This book has focused on eco-justice activism during the COVID-19 pandemic, addressing its policing and social control, as well as the transformative potential borne by its visual and performative practices of resistance. Drawing on an extensive ethnography of two eco-justice movements in the northern Italian city of Trento, this book has shown how the policing of eco-justice protest intensified during the pandemic, involving the deployment of larger numbers of police personnel and containment measures – the latter mostly to protect buildings symbolizing power and soon-to-be construction sites operated by private companies. Protest policing practices predicated on the maximizing of police visibility are in themselves not new: they have also been used by the police to police 'transgressive' protest before the pandemic. However, they were never used with that intensity against the two movements under study in spite of them being considered problematic by the police long before the pandemic began. Protest policing strategies like the ones described in this book help construct protesters as 'enemies' to be feared and legitimize the enhanced police control of activists, with their reach potentially extending to the pandemic aftermath.

The book has also demonstrated how, in Trento, governing eco-justice protest involved making activists and their grievances less visible to the public eye, including by displacing protesters outside the city centre. The book argued that such displacement mostly had an economic reasoning underpinning it: that of protecting local businesses from unwelcome protest-related disruptions. Such reasoning is not unique to Trento's public order governance. As mentioned earlier in this book, other cities in Italy and Europe, and even whole countries, have also recently tightened their regulations on public protesting to safeguard the local economy from potentially disrupting mobilizations.

Even when their protests are displaced outside the urban centre, activists can always find ways to keep their grievances in public sight. For example, activists can march in the city centre in spite of the denied police permission, as the SC campaign did during my fieldwork (see Chapter One). Activists can also use social media to convey their otherwise-unheard messages (see, for example, Di Ronco et al, 2019; Ismangil and Lee, 2021) and populate the city with markers of visual resistance, including flags, graffiti, stickers and stencils, among other things (see Chapter Three). All these activist practices contribute to offering a glimpse into alternatives to the status quo – a window facing scenarios where human progress does not come at the cost of ecological degradation and destruction.

With its thorough analysis of the policing of eco-justice protest, this book contributes to green critical criminology and, in particular, to the critical criminological scholarship on the social control and penalization of activism. Specifically, Chapter One of this the book highlights continuity with the previous critical criminological literature that has observed increasing authoritarian attitudes towards social movements and their mobilizations in recent years (Maroto et al, 2019; Vegh Weiss, 2021a). Such attitudes were already observed by critical scholars during the 2010s (Maroto et al, 2019), as well as more recently during large and petty 'states of exception' (Fritsch and Kretschmann, 2021), which often coincided with political, economic, social and health crises (Burnett et al, 2022). Focusing on the case study of the policing of eco-justice movements in Trento during the COVID-19 pandemic, this book provided empirical evidence supporting this critical scholarship and has demonstrated how police attitudes towards eco-justice protest worsened during the COVID-19 health crisis.

In addition, in line with Maroto, González-Sánchez and Brandariz (2019), this study noted the importance of administrative measures (in addition to the criminal law) in the penalization of social movements and their protesting. During the pandemic, these measures have included administrative fines (as in Maroto et al's [2019] study) and, more generally, stricter regulations and conditions on protesting, in addition to increased militarized police practices to policing public protest (Martin, 2021, 2022).

This book also contributes to the multidisciplinary and critical criminological literatures on the governance of urban space, which critically analyse the exclusion of the most vulnerable – and increasingly also of activists – from consumption-focused city centres (see Chapter Two). Crucially, the book conceptualizes urban space as a battleground where different constructions of space, everyday meanings and expectations coexist and often come into conflict. Values, norms and expectations that are not aligned with 'dominant' ones are often expressed through visual and performative practices, through which activists, among others, challenge hegemonic social orders and open the door to alternative imaginaries of urban living. As Chapter Three of this book demonstrated, these practices often convey knowledge of green harms and invite a more respectful attitude to the more-than-human surrounding us. The senses and affect are essential conveyors of such knowledge and imaginaries, and should therefore be more consistently attended to by researchers while engaging in qualitative green critical criminological analyses, as recent critical sensory criminology has also suggested (McClanahan and South, 2020).

The book has attended to both power and resistance, and it is precisely here that its key strength lies: it complements a critical analysis of expanding state control on eco-justice activists during the pandemic with reflections on the radical possibilities that activist practices often reveal to us, shaping the way we understand urban space and behave in it. Such integrated analyses are rare in critical criminology. Indeed, as Ruggiero (2003, 2021b) reminds us, critical criminology (with the exception of its early days) has mostly focused on the analysis of criminalization and marginalization processes, rather than on conflict and social action leading to social change.

Focusing on activist radical imaginaries

Regular research is certainly needed to monitor and critically analyse the policing and criminalization of dissent and protest after the pandemic and during the next political, economic, social or health crisis. Critical (green) criminologists will surely keep doing that, exposing how state actors (and economic and political systems more generally) cause destruction, devastation and suffering at the intersection of race, ethnicity, class, gender, sex, age and species, among other things. My suggestion is that such analyses also combine multilayered accounts of resistance or 'disobedience' (Pali, 2022) on the understanding that the doings of power are always relational and never unilaterally imposed on people and silently adhered to by them. In particular, by also focusing on activist visual and performative practices, critical criminologists can contribute to both exposing and amplifying activist knowledge and the radical imaginaries their practices often glimpse at, ultimately contributing to social change. When published in international journals, presented at international conferences and talked about in public outlets and at public events, this type of critical criminological research can seek to obtain at least two goals: first, it can encourage members of the public to acknowledge officially untold or unrecognized harms, rousing them to act (Cohen, 2001); and, second, it can facilitate the entrance of activist knowledge and imaginaries into the public imagination, ultimately shaping policy and practice, while, at the same time, also 'inaugurat[ing] a process of imagining and living more progressive alternatives' (Ferrell, 2022: 607; see also Kramer, 2016).

In essence, my suggestion and invitation for critical criminology is to conduct research that assists activists – never guides or leads them (see Naegler, 2021; see also the Introduction to this book) – through exploring, exposing and amplifying their voices, critical knowledge and radical imaginaries for the future. This also means trying to move beyond critical analyses that only focus on activists' oppression, which are publicly disseminated by scholars to the public in several ways. Similar to other critical criminologists (see, for example, Kramer, 2016; Ozymy et al, 2020), I find this an essential first step; however, something more is needed to enrich critical criminological analyses and public engagement efforts in ways that are truly supportive of activists' political endeavours, needs and demands.[1] That something is an attention to activist practices and their suggested radical imaginaries.

As I argued in Chapter Three using the language of atmospheres, a way for green critical criminologists to do so is through designing participatory research projects that help capture the ways in which activists and the public understand resistance and contestation, as well as, most importantly, how they imagine future scenarios for social and ecological change. Such projects can also involve or intentionally be set in spaces that are threatened or are affected by ecological destruction (for example, spaces where ecosystems, plants or animals have been, are or will soon be harmed by humans), thus opening up new radical possibilities for such spaces. In Chapter Three, I also suggested that green critical criminological research engages more regularly with interdisciplinary collaborations with, among other things, the arts and humanities, which are disciplines that have proven successful in assisting humans to 'be with' the more-than-human and understand their suffering. Such collaborations are key to capturing the harms suffered by the more-than-human surrounding us and to feeding this knowledge into ideas of repair, conservation and sustainability for our future living.

My suggestions for future (green) critical criminological scholarship are compatible with the position known as 'activist critical criminology', which motivates critical criminologists to put their research and teaching at the service of social and legal justice (Belknap, 2015) and those victimized (Goyes, 2016). Such a position is inspired by the need to break through state denial of harms (Kramer, 2016), affect change, tackle injustices and prevent victimization and harm (see Arrigo, 2016). Indeed, by exploring and exposing activists' radical alternatives, critical criminologists can go a long way to facilitating the public's acknowledgement of harms, as well as triggering people's indignation, reducing suffering and preventing harm and ecological destruction.

My suggestions are also compatible with the notion of (nomocentric) 'troublemaker' put forward by Ruggiero (2021b), who used it recently as a theme to try to revive critical criminology. In particular, Ruggiero used this theme to encourage critical criminologists to actively engage with social movements in an effort to fuel social transformation. In agreement with Ruggiero, I believe that transformative critical criminology can only happen when critical criminologists actively engage with social movements and fight alongside them to achieve change. As I have argued in these concluding thoughts, fighting alongside activists also means valorizing their critical knowledge through research that exposes and amplifies their ideas and imaginaries for our future living. Ultimately, such research can also help us understand the conditions that foster and sustain cultures of resistance and disobedience (Pali, 2022), which are so important to counteracting harmful exercises of power and reducing suffering on our planet.

Notes

Introduction

1. See: https://lab24.ilsole24ore.com/qualita-della-vita/trento (accessed 24 February 2023).
2. For the website of the NOTAV Committee in Trento, see: www.notavtrentino.it/ (accessed 24 February 2023).
3. See: https://transport.ec.europa.eu/transport-themes/infrastructure-and-investment/trans-european-transport-network-ten-t_en (accessed 24 February 2023).
4. See: www.dpcirconvallazioneferroviariatrento.it/inquadramento/ (accessed 24 February 2023).
5. See: www.governo.it/sites/governo.it/files/PNRR.pdf (accessed 24 February 2023).
6. See: https://ec.europa.eu/info/strategy/recovery-plan-europe_en (accessed 24 February 2023).
7. For the website of the SC campaign, see: https://assembleantispecista.noblogs.org/stopcasteller/ (accessed 1 March 2023).
8. For the website of the Bruno social centre, see: https://csbruno.org/ (accessed 24 February 2023).
9. See: https://assembleantispecista.noblogs.org/stopcasteller/ (accessed 24 February 2023). The video where the poor conditions of the bears are exposed is also available on this webpage – in the 'video' section.

one Flexing the Muscles of Power: Policing Urban Eco-Justice Activism During the Pandemic

1. In Italy, such an approach has mainly been codified through internal memos by the Interior Ministry and its Public Safety Department, as well as by police manuals and the School of Public Order – not, strictly speaking, by law, thus leaving the police with wide operational discretion and little accountability (Gargiulo, 2015; Fabini and Sbraccia, 2021; Tuzza, 2021).
2. This contravention is envisaged by Article 18 of the Testo Unico delle Leggi di Pubblica Sicurezza or TULPS (Royal Decree No. 773 of 18 June 1931).
3. The possibility to ban 'uncivil' people has been given to mayors by two 'security decrees': law decree No 14 of 20 February 2017 and law decree No 113 of 4 October 2018. The second decree is particularly important as

NOTES

it specifically targeted dissenters in various ways (see Selmini, 2020). For example, it converted some administrative offences (such as that of traffic block) into crime proper and punished them through increased penalties. It also aggravated penalties for existing criminal offences, including for damaging property during a public protest and for threatening and resisting the police (which are offences now punishable with up to and at least 5 years imprisonment, respectively).

[4] There are 44 infractions in Section 2 of the 2015 Ley Orgánica de Protección de la Seguridad Ciudadana (or LOPSC), which distinguishes between very serious, serious and mild violations. For these infractions and their associated administrative fines, see the Ley Orgánica 4/2015 of 30 March 2015, available at: www.boe.es/boe/dias/2015/03/31/pdfs/BOE-A-2015-3442.pdf (accessed 24 February 2023).

[5] See the DPCM of 14 January 2021 (which is relevant for the present study) available from: www.governo.it/sites/new.governo.it/files/Dpcm_14_gennaio_2021.pdf (accessed 24 February 2023). The requirement of static protest was also in place during the year 2020, see, for example, the DPCM of 17 May 2020, available from: www.governo.it/sites/new.governo.it/files/DPCM_20200517.pdf (accessed 24 February 2023).

[6] Attended in-person events were held on: 2 December 2020 (NOTAV); 9 February 2021 (NOTAV); 13 February 2021 (NOTAV); 8 May 2021 (NOTAV); 15 May 2021 (NOTAV); 21 May 2021 (NOTAV); 1 June 2021 (NOTAV); 5 June 2021 (NOTAV); 5 June 2021 (SC); 11 June 2021 (NOTAV); 15 June 2021 (NOTAV); 22 June 2021 (NOTAV); 26 June 2021 (NOTAV and SC); 3 July 2021 (NOTAV); and 6 June 2021 (NOTAV).

[7] Online events were held on: 18 February 2021 (NOTAV); 7 March 2021 (SC); 9 March 2021 (NOTAV); 16 March 2021 (NOTAV); 23 March 2021 (NOTAV); 30 March 2021 (NOTAV); 10 April 2021 (SC); and 9 June 2021 (NOTAV).

[8] In the past, some anarchists in Trento were convicted for damaging public properties, and in 2013, some of them were even trialled on charges of domestic terrorism (all of them were acquitted) (see, for example, TGR Trento, 2019).

[9] See also: www.instagram.com/p/CNfePH1FFCi/ (accessed 24 February 2023).

[10] See, for example: www.giornaletrentino.it/cronaca/trento/no-tav-blitz-anarchico-sulla-trivella-a-mattarello-1.1134269) (accessed 24 February 2023). According to activists, back then, survey sites were subject to little police surveillance.

[11] The pass was introduced in August 2021 and (until April 2022) was mandatory to dine indoors, visit museums and theatres, and (from October 2021 until April 2022) enter all workplaces.

two Power, Consumption, Disorder and Protest in Inner-City Centres

[1] Regeneration has also affected urban areas that are in the geographical proximity of city centres, including those that were previously dilapidated (Smith, 1996). In these cases, the strategy of branding has been found to be important in changing the image of the neighbourhood and opening the doors to further gentrification. For some examples of gentrification in previously dilapidated 'no-go' areas, see Naegler (2012) and Rius Ulldemolins (2014).

[2] Imprisonment would not exceed six months or five years, depending on whether the conviction is summary or on indictment, respectively.

[3] See: https://manifestoclub.info/about/ (accessed 24 February 2023).

[4] For the list of passed PSPOs and their targeted activities as per 2016, see: https://manifestoclub.info/psposreport/ (accessed 24 February 2023).

[5] The possibility to ban 'uncivil' people was given to mayors by two 'security decrees': the so-called 'Minniti decree' (Law Decree No. 14 of 20 February 2017) and the first so-called 'Salvini decree' (Law Decree No. 113 of 4 October 2018). Both these decrees borrow their name from their proponents, that is, the then Ministers of the Interior Marco Minniti and Matteo Salvini, respectively.

[6] The length of the ban and of the associated sanction for breach are higher for people who have previously been found guilty of certain crimes. In practice, the criminal sanction has almost never been applied (see Borlizzi, 2022).

[7] In their article, Bannister et al (2006) analysed the UK New Labour's approach to incivility in the city in the 1990s and early 2000s. While also using Neil Smith's (1996) idea of gentrification as revanchism, they argued that the regenerated 'respectable' city is also a revanchist city: a city cleansed of difference through vengeful punitive regulations and zero-tolerance policing against incivilities. For a more tentative application of Smith's theory in the UK context, see Atkinson (2003).

[8] See: www.antwerpen.be/nl/overzicht/stadsorganisatie-1/detail/politiecodex (accessed 24 February 2023).

[9] A number of Just Stop Oil activists have recently been charged with causing public nuisance, or with conspiracy to cause public nuisance, in relation to disruptive protests that occurred on the M25 motorway in November 2022. During such protests, activists climbed onto gantries at various locations over London's M25 motorway, stopping traffic in both directions (see Gayle, 2022b).

[10] The rejected provisions included: the creation of 'locking on' offences and serious disruption prevention orders; the criminalization of obstructions

of major transport works and interference with the operation of key national infrastructure; and the extension of police powers to stop and search without suspicion of an offence (see UK Parliament, 2022a).

three Atmospheres of Eco-Justice Resistance During the Pandemic

1. Especially in the early months of fieldwork when face-to-face encounters were restricted due to COVID-19 regulations, I spent time walking around the city and photographing markers of visual eco-justice resistance as they appeared. They included: NOTAV tags, flags, stickers, stencils and banners; XR stickers; one 'on the side of nature' tag in a large outdoor parking lot; NOTAP tags and stickers; 'free bears' banners; 'missing' stickers referring to wild bears killed by humans; #stopcasteller stickers; a 'freedom for M45' (an imprisoned bear) stencil; bear footprints on the pavement; and eco-justice protest flyers (both the days before and on the day of demonstrations). In general, during fieldwork, and in particular during protests, I found it very hard to take pictures of the police given their intimidating presence and levels of surveillance of activists (and hence myself). For this reason, the part of this chapter focused on the rupture of the protest's atmosphere due to police presence does not include any pictures of the police.
2. For a video summarizing the event, see: https://notavtrento.noblogs.org/post/2021/05/20/15-maggio-a-san-martino-visita-a-un-cantiere-da-evitare-radio-s-martino-tav/?fbclid=IwAR0z-LmfnSib_CRVBGC98mjiMOfzMY8GZOrjhAFKW06sirk9XJaUSEkDRw8 (accessed 24 February 2023).
3. Criminalized movements often engage in sousveillance and the counter-surveillance of the police (see, for example, Di Ronco and Allen-Robertson, 2021).
4. For another interesting interdisciplinary project that also reclaimed a more emphatic relationship with rivers through walking and other art-based practices, see: https://entre-rios.net/ (accessed 24 February 2023). For an overview of this project in English, see: www.youtube.com/watch?v=Kh9dNRPZjq8&feature=youtu.be (accessed 24 February 2023).

Conclusion

1. Let us not forget that many social movements are often well aware of the political economy of oppression and themselves produce rich critical knowledge on, among other things, police practices and repression by engaging with other criminalized movements.

References

Agnew, R. (2020) 'The ordinary acts that contribute to ecocide: a criminological analysis', in A. Brisman and N. South (eds) *Routledge International Handbook of Green Criminology* (2nd edn), London: Routledge, pp 52–67.

Amin, A. and Thrift, N. (2017) *Seeing like a City*, London: Polity Press.

Anderson, B. (2009) 'Affective atmospheres', *Emotion, Space and Society*, 2(2): 77–81.

Anderson, B. and Ash, J. (2015) 'Atmospheric methods', in P. Vannini (ed) *Non-representational Methodologies: Re-envisioning Research*, London: Routledge, pp 34–51.

Andron, S. (2018) 'Selling streetness as experience: the role of street art tours in branding the creative city', *The Sociological Review*, 66(5): 1036–57.

Arenas, I. and Sweet, E.L. (2019a) 'Disassembling cities: spatial, social and conceptual trajectories across the urban globe', in E.L. Sweet (ed) *Disassembled Cities: Social and Spatial Strategies to Reassemble Communities*, London: Routledge, pp 3–14.

Arenas, I. and Sweet, E.L. (2019b) 'The organizing logics of predatory formations: militarization and the spectacle of the (in)security state', in E.L. Sweet (ed) *Disassembled Cities: Social and Spatial Strategies to Reassemble Communities*, London: Routledge, pp 73–82.

Ares, E. and Bolton, P. (2020) 'The rise of climate change activism?', House of Commons Library, 24 June. Available from: https://commonslibrary.parliament.uk/the-rise-of-climate-change-activism/ (accessed 4 January 2023).

Arrigo, B.A. (2016) 'Critical criminology as academic activism: on praxis and pedagogy, resistance and revolution', *Critical Criminology*, 24(4): 469–71.

Ashworth, A. (2004) 'Social control and "anti-social behaviour": the subversion of human rights', *Law Quarterly Review*, 120(2): 263–91.

Ashworth, A. and Zedner, L. (2014) *Preventive Justice*, Oxford: Oxford University Press.

REFERENCES

Atkinson, R. (2003) 'Domestication by cappuccino or a revenge on urban space? Control and empowerment in the management of public spaces', *Urban Studies*, 40(9): 1829–43.

Bannister, J., Fyfe, N. and Kearns, A. (2006) 'Respectable or respectful? (In)Civility and the city', *Urban Studies*, 43(5–6): 919–37.

Bauman, Z. (1997) *Postmodernity and its Discontents*, Cambridge: Polity Press.

BBC (British Broadcasting Corporation) (2022) 'Freedom Convoy: Paris protest banned by police ahead of arrival', 10 February. Available from: www.bbc.com/news/world-europe-60317807 (accessed 8 July 2022).

Belknap, J. (2015) 'Activist criminology: criminologists' responsibility to advocate for social and legal justice', *Criminology*, 53(1): 1–22.

Bergamaschi, M., Castrignanò, M. and Rubertis, P.D. (2014) 'The homeless and public space: urban policy and exclusion in Bologna', *Revue Interventions Économiques. Papers in Political Economy*, 51. Available from: https://doi.org/10.4000/interventionseconomiques.2441 (accessed 24 February 2023).

Bombardi, L. and Porto Almeida, V. (2022) 'Amazon under siege: an interview with environmental and human rights defender Claudelice dos Santos', *Criminological Encounters*, 5(1). Available from: http://doi.org/10.26395/CE22050117 (accessed 24 February 2023).

Borlizzi, F. (2022) 'Daspo urbano e governo delle città: riflessioni a margine di una ricerca empirica', *Studi sulla Questione Criminale*, 3 November. Available from: https://studiquestionecriminale.wordpress.com/2022/11/03/daspo-urbano-e-governo-delle-citta-riflessioni-a-margine-di-una-ricerca-empirica/?fbclid=IwAR1wnBzJXPeH9gbjP2CsWmVIl_-XdzHz3LyorUIWJBh4_atCAYA3KJ2xhPw (accessed 5 January 2023).

Brisman, A. and South, N. (2013) 'A green-cultural criminology: an exploratory outline', *Crime, Media, Culture*, 9(2): 115–35.

Brisman, A. and South, N. (2014) *Green Cultural Criminology: Constructions of Environmental Harm, Consumerism, and Resistance to Ecocide*, London and New York, NY: Routledge.

Brisman, A. and South, N. (2020) 'The growth of a field: a short history of a "green" criminology', in A. Brisman and N. South (eds) *Routledge International Handbook of Green Criminology* (2nd edn), London and New York, NY: Routledge, pp 39–51.

Brisman, A., García Ruiz, A., McClanahan, B. and South, N. (2021) 'Exploring sound and noise in the urban environment: tensions between cultural expression and municipal control, health and inequality, police power and resistance', in N. Peršak and A. Di Ronco (eds) *Harm and Disorder in the Urban Space: Social Control, Sense and Sensibility*, London: Routledge, pp 15–29.

Brock, A. and Goodey, J. (2022) 'Policing the High Speed 2 (HS2): repression and collusion along Europe's biggest infrastructure project', in A. Dunlap and A. Brock (eds) *Enforcing Ecocyde: Power, Policing and Planetary Militarization*, Cham: Palgrave Macmillan, pp 227–68.

Brown, J., Kirk-Wade, E., Baker, C. and Barber, S. (2021) 'Coronavirus: a history of English lockdown laws', House of Commons Library, 22 December. Available from: https://commonslibrary.parliament.uk/research-briefings/cbp-9068/ (accessed 4 January 2023).

Brown, K.J. (2017) 'The hyper-regulation of public space: the use and abuse of public spaces protection orders in England and Wales', *Legal Studies*, 37(3): 543–68.

Brown, M. and Carrabine, E. (2017) *Routledge International Handbook of Visual Criminology*, London: Routledge.

Brown, A., Parrish, W. and Speri, A (2017) 'Leaked documents reveal counterterrorism tactics used at Standing Rock to "Defeat Pipeline Insurgencies"', *The Intercept*, 27 May. Available from: https://theintercept.com/2017/05/27/leaked-documents-reveal-security-firms-counterterrorism-tactics-at-standing-rock-to-defeat-pipeline-insurgencies/ (accessed 1 March 2023).

Burnett, J., Nafstad, I. and White, L. (2022) 'Introduction to the special issue: pandemics, policing and protest', *Justice, Power and Resistance*, 5(1–2): 2–8.

Burney, E. (2005) *Making People Behave: Anti-social Behaviour, Politics and Policy*, Cullompton: Willan Publishing.

REFERENCES

Burney, E. (2008) 'The ASBO and the shift to punishment', in P. Squires (ed) *ASBO Nation: The Criminalisation of Nuisance*, Bristol: The Policy Press, pp 135–48.

Campbell, E. (2013) 'Transgression, affect and performance: choreographing a politics of urban space', *British Journal of Criminology*, 53(1): 18–40.

Chabrol, M., Collet, A., Giroud, M., Launay, L., Rousseau, M. and Minassian, H.T. (2022) *Gentrifications: Views from Europe*, New York, NY, and Oxford: Berghahn.

Chiaramonte, X. (2019) *Governare il Conflitto: la Criminalizzazione del Movimento No TAV*, Milan: Mimesis.

Cohen, S. (2001) *States of Denial: Knowing about Atrocities and Suffering*, Cambridge: Polity Press.

Comune Catania (2017) 'Firmata l'ordinanaza, multe a prostitute su strada e clienti'. Available from: www.comune.catania.it/infor mazioni/cstampa/default.aspx?cs=58823 (accessed 24 June 2022).

Coordinamento Trentino NOTAV (2019) 'TAV in Trentino: un buco nell'acqua. Tutto quello che non ti aspetti', *Comitato No Tav Trento*, 17 January. Available from: https://notavtrento.noblogs. org/post/2019/01/17/scarica-il-nuovo-opuscolo-tav-in-trent ino-un-buco-nellacqua/ (accessed 8 July 2022).

Coordinamento Trentino NOTAV (2021) 'Coord. NO TAV Circonvallazione di Trento e il raddoppio della ferrovia in Trentino: sogno o incubo?', YouTube, 31 March. Available from: www.youtube.com/watch?v=cgw6iIIgSVk (accessed 22 June 2022).

Crawford, A. (2009) 'Governing through anti-social behaviour: regulatory challenges to criminal justice', *The British Journal of Criminology*, 49(6): 810–31.

Crawford, A. and Evans, K. (2017) 'Crime prevention and community safety', in A. Liebling, S. Maruna and L. McAra (eds) *The Oxford Handbook of Criminology* (6th edn), Oxford: Oxford University Press, pp 797–824.

Crocitti, S. and Selmini, R. (2017) 'Controlling immigrants: the latent function of Italian administrative orders', *European Journal on Criminal Policy and Research*, 23(1): 99–114.

Crosby, A. and Monaghan, J. (2018) *Policing Indigenous Movements: Dissent and the Security State*, Halifax and Winnipeg: Fernwood Publishing.

CSA Bruno (2021a) 'Circonvallazione ferroviaria: troppo poco, troppo tardi', *CiclOstyle*, 24 February. Available from: https://ciclostile.csbruno.org/2021/02/24/circo nvallazione-ferroviaria-troppo-poco-troppo-tardi/?fbclid= IwAR0NYrjA_j0zj9MHmGSn6s7HzoXcpzixpPBbSo_I xYC_rKPLGvDV9EVCVec (accessed 8 July 2022).

CSA Bruno (2021b) 'Stop Casteller: 7 compagni accusati di danneggiamento mentre l'orso M57 viene trasferito in uno zoo', 23 December. Available from: https://csbruno.org/stop-castel ler-7-compagni-accusati-di-danneggiamento-mentre-lorso-m57- viene-trasferito-in-uno-zoo/ (accessed 8 July 2022).

Davidson, M. and Lees, L. (2010) 'New-build gentrification: its histories, trajectories, and critical geographies', *Population, Space and Place*, 16(5): 395–411.

Davis, M. (1990) *City of Quartz: Excavating the Future in Los Angeles*, London and New York, NY: Verso.

De Jong, A. and Schuilenburg, M. (2006) *Mediapolis: Popular Culture and the City*, Rotterdam: 010 Publishers.

della Porta, D. (1998) 'Police knowledge and public order: some reflections on the Italian case', in D. della Porta and H. Reiter (eds) *Policing Protest: The Control of Mass Demonstrations in Western Democracies*, Minneapolis, MN: University of Minnesota Press, pp 228–52.

della Porta, D. and Fillieule, O. (2004) 'Policing social protest', in D.A. Snow, S.A. Soule and H. Kriesi (eds) *The Blackwell Companion to Social Movements*, Oxford: Blackwell Publishing, pp 217–41.

della Porta, D. and Reiter, H. (1998) *Policing Protest: The Control of Mass Demonstrations in Western Democracies*, Minneapolis, MN, and London: University of Minnesota Press.

della Porta, D. and Reiter, H. (2003) *Polizia e Ordine Pubblico*, Bologna: Il Mulino.

della Porta, D. and Zamponi, L. (2013) 'Protest and policing on October 15th, global day of action: the Italian case', *Policing and Society*, 23(1): 65–80.

REFERENCES

Dickinson, S., Millie, A. and Peters, E. (2022) 'Street skateboarding and the aesthetic order of public spaces', *British Journal of Criminology*, 62(6): 1454–69.

Di Ronco, A. (2016) 'Inspecting the European crime prevention strategy towards incivilities', *Crime Prevention and Community Safety*, 18(2): 14–160.

Di Ronco, A. (2018) 'Disorderly or simply ugly? Representations of the local regulation of street prostitution in the Italian press and their policy implications', *International Journal of Law, Crime and Justice*, 52, 10–22.

Di Ronco, A. (2021a) 'Power at play: the policing of sex work in two European cities', in N. Peršak and A. Di Ronco (eds) *Harm and Disorder in the Urban Space: Social Control, Sense and Sensibility*, London: Routledge, pp 142–64.

Di Ronco, A. (2021b) 'What happened when Italy criminalised environmental protest', *The Conversation*, 12 May. Available from: https://theconversation.com/what-happened-when-italy-criminalised-environmental-protest-158014 (accessed 20 June 2022).

Di Ronco, A. (2022) 'Law in action: local-level prostitution policies and practices and their effects on sex workers', *European Journal of Criminology*, 19(5): 1078–96.

Di Ronco, A. and Allen-Robertson, J. (2021) 'Representations of environmental protest on the ground and in the cloud: the NOTAP protests in activist practice and social visual media', *Crime, Media, Culture*, 17(3): 375–99.

Di Ronco, A. and Chiaramonte, X. (2022) 'Harm to knowledge: criminalising environmental movements speaking up against megaprojects', in B. Pali, M. Fosyth and F. Tepper (eds) *The Palgrave Handbook of Environmental Restorative Justice*, Cham: Palgrave Macmillan, pp 421–47.

Di Ronco, A. and Peršak, N. (2014) 'Regulation of incivilities in the UK, Italy and Belgium: courts as potential safeguards against legislative vagueness and excessive use of penalising powers?', *International Journal of Law, Crime and Justice*, 42(4): 340–65.

Di Ronco, A. and Sergi, A. (2019) 'From harmless incivilities to not-so serious organised crime activities: the expanded realm of European crime prevention and some suggestions on how to limit it', in P.C. van Duyne, A. Serdyuk, G.A. Antonopoulos, J.H. Harvey and K. von Lampe (eds) *Constructing and Organising Crime in Europe*, The Hague: Eleven International Publishing, pp 149–78.

Di Ronco, A., Allen-Robertson, J. and South, N. (2019) 'Representing environmental harm and resistance on Twitter: the case of the TAP pipeline in Italy', *Crime, Media, Culture*, 15(1): 143–68.

Ellefsen, R. (2018) 'Relational dynamics of protest and protest policing: strategic interaction and the coevolution of targeting strategies', *Policing and Society*, 28(7): 751–67.

Elliott-Cooper, A., Hubbard, P. and Lees, L. (2020) 'Moving beyond Marcuse: gentrification, displacement and the violence of unhoming', *Progress in Human Geography*, 44(3): 492–509.

Fabini, G. and Sbraccia, A. (2021) 'Criminal policies in action: Italian police forces, discretionary powers, and selective law enforcement', in J.M. Mbuba (ed) *Global Perspectives in Policing and Law Enforcement*, Lanham, MD: Lexington Books, pp 99–114.

Fatsis, L. (2019) 'Grime: criminal subculture or public counterculture? A critical investigation into the criminalization of Black musical subcultures in the UK', *Crime, Media, Culture*, 15(3): 447–61.

Fatsis, L. and Lamb, M. (2021) *Policing the Pandemic: How Public Health Becomes Public Order*, Bristol: Bristol University Press.

Ferree, M.M. (2004) 'Soft repression: ridicule, stigma, and silencing in gender-based movements', in D.J. Myers and D.M. Cress (eds) *Authority in Contention*, Bingley: Emerald Group Publishing Limited, pp 85–101.

Ferrell, J. (1997) 'Youth, crime, and cultural space', *Social Justice*, 24(4): 21–38.

Ferrell, J. (2022) 'In defense of resistance', *Critical Criminology*, 30: 603–619.

Finnegan, W. (2020) 'Environmental activism goes digital in lockdown – but could it change the movement for good?', *The Conversation*, 7 May. Available from: https://theconversation.com/environmental-activism-goes-digital-in-lockdown-but-could-it-change-the-movement-for-good-137203 (accessed 20 June 2022).

REFERENCES

Florida, R. (2002) *The Rise of the Creative Class*, New York, NY: Basic Books.

Foucault, M. (1977) *Discipline and Punish: The Birth of the Prison*, London: Allen Lane.

France24 (2022) 'Ottawa "Freedom Convoy" forces factory shutdowns as Trudeau condemns "unacceptable" tactics', 10 February. Available from: www.france24.com/en/americas/20220210-canadian-covid-19-blockade-forces-auto-factory-shutdowns-as-trudeau-slams-unacceptable-tactics (accessed 20 June 2022).

Fraser, A. (2021) 'The street as an affective atmosphere', in K. Herrity, B.E. Schmidt and J. Warr (eds) *Sensory Penalties: Exploring the Senses in Spaces of Punishment and Social Control*, Bingley: Emerald Publishing Limited, pp 217–30.

Fraser, A. and Matthews, D. (2021) 'Towards a criminology of atmospheres: law, affect and the codes of the street', *Criminology & Criminal Justice*, 21(4): 455–71.

Fritsch, K. and Kretschmann, A. (2021) 'Politics of exception: criminalising activism in Western democracies', in V. Vegh Weiss (ed) *Criminalization of Activism: Historical, Present and Future Perspectives*, London: Routledge, pp 19–29.

Garcia Ruiz, A. and South, N. (2019) 'Surrounded by sound: noise, rights and environments', *Crime, Media, Culture*, 15(1): 125–41.

Gargiulo, E. (2015) 'Ordine pubblico, regole private. Rappresentazioni della folla e prescrizioni comportamentali nei manuali per i reparti mobili', *Etnografia e Ricerca Qualitativa*, 8(3): 481–512.

Gayet-Viaud, C. (2017) 'French cities' struggle against incivilities: from theory to practices in regulating urban public space', *European Journal on Criminal Policy and Research*, 23(1): 77–97.

Gayle, D. (2022a) 'Just Stop Oil says only threat of death sentence would stop its protests', *The Guardian*, 21 October. Available from: www.theguardian.com/environment/2022/oct/21/just-stop-oil-says-only-threat-of-death-sentence-would-stop-its-protests (accessed 1 November 2022).

Gayle, D. (2022b) 'More than 30 climate activists behind bars in UK during Cop27', *The Guardian*, 21 November. Available from: www.theguardian.com/environment/2022/nov/21/more-than-30-climate-activists-just-stop-oil-were-behind-bars-in-uk-during-cop27 (accessed 5 January 2023).

Gillham, P. and Noakes, J. (2007) '"More than a march in a circle": transgressive protests and the limits of negotiated management', *Mobilization: An International Quarterly*, 12(4): 341–57.

Gilmore, J., Jackson, W., Monk, H. and Short, D. (2020) 'Policing the UK's anti-fracking movement: facilitating peaceful protest or facilitating the industry?', *Peace Human Rights Governance*, 4(3): 349–90.

Global Project (2021) 'Trento-manganellate contro manifestanti antispecisti', YouTube, 10 April. Available from: www.youtube.com/watch?v=_hGLS97ZHkk&list=PLdvH4tPHvBJ_k27g5Zd1u-elOX5FYEJLb (accessed 24 June 2022).

Global Witness (2021) 'Last line of defence'. Available from: www.globalwitness.org/en/campaigns/environmental-activists/last-line-defence/ (accessed 24 June 2022).

Gobby, J. and Everett, L. (2022) 'Policing Indigenous land defense and climate activism: learnings from the frontlines of pipeline resistance in Canada', in A. Dunlap and A. Brock (eds) *Enforcing Ecocyde: Power, Policing and Planetary Militarization*, Cham: Palgrave Macmillan, pp 89–121.

Governo Italiano (2021) 'Coronavirus, le misure adottate dal governo'. Available from: www.sitiarcheologici.palazzochigi.it/www.governo.it/febbraio%202021/it/coronavirus-misure-del-governo.html (accessed 21 June 2022).

Governo Italiano (2022) 'Coronavirus, le misure adottate dal governo'. Available from: www.governo.it/it/coronavirus-misure-del-governo (accessed 21 June 2022).

Gov.uk (2022) 'Protest powers: Police, Crime, Sentencing and Courts Act 2022 factsheet'. Available from: www.gov.uk/government/publications/police-crime-sentencing-and-courts-bill-2021-factsheets/police-crime-sentencing-and-courts-bill-2021-protest-powers-factsheet (accessed 21 June 2022).

REFERENCES

Goyes, D.R. (2016) 'Green activist criminology and the epistemologies of the South', *Critical Criminology*, 24(4): 503–18.

Grottolo, T. (2021) 'Tram, ascensori verticali e tanto "green" (VIDEO), ecco la Trento del futuro immaginata dopo la Circonvallazione ferroviaria', *Il Dolomiti*, 5 November. Available from: www.ildolomiti.it/societa/2021/tram-ascensori-verticali-e-tanto-green-video-ecco-la-trento-del-futuro-immaginata-dopo-la-circonvallazione-ferroviaria (accessed 20 June 2022).

Harvey, D. (1989) 'From managerialism to entrepreneurialism: the transformation in urban governance in late capitalism', *Geografiska Annaler: Series B, Human Geography*, 71(1): 3–17.

Hasler, O., Walters, R. and White, R. (2020) 'In and against the state: the dynamics of environmental activism', *Critical Criminology*, 28(3): 517–31.

Hayward, K.J. (2004) *City Limits: Crime, Consumer Culture and the Urban Experience*, London, Sydney and Portland, OR: Glass House Press.

Hayward, K.J. (2012) 'Five spaces of cultural criminology', *The British Journal of Criminology*, 52(3): 441–62.

Hayward, K.J. and Schuilenburg, M. (2014) 'To resist = to create? Some thoughts on the concept of resistance in cultural criminology', *Tijdschrift Over Cultuur & Criminaliteit*, 4(1): 22–36.

Hayward, K.J. and Yar, M. (2006) 'The "chav" phenomenon: consumption, media and the construction of a new underclass', *Crime, Media, Culture*, 2(1): 9–28.

Herrity, K., Schmidt, B.E. and Warr, J. (2021) *Sensory Penalties: Exploring the Senses in Spaces of Punishment and Social Control*, Bingley: Emerald Publishing Limited.

HMICFRS (Her Majesty's Inspectorate of Constabulary and Fire & Rescue Services) (2021) 'Getting the balance right? An inspection of how effectively the police deal with protests', 11 March. Available from: www.justiceinspectorates.gov.uk/hmicfrs/publications/getting-the-balance-right-an-inspection-of-how-effectively-the-police-deal-with-protests/ (accessed 20 June 2022).

Holloway, L. and Hubbard, P. (2001) *People and Place: The Extraordinary Geographies of Everyday Life*, London and New York, NY: Routledge.

Interno (2021) 'Direttiva del ministro Lamorgese per lo svolgimento di manifestazioni di protesta contro le misure sanitarie'. Available from: www.interno.gov.it/sites/default/files/2021-11/dirett iva_del_ministro_10-11-2021.pdf (accessed 24 June 2022).

Ismangil, M. and Lee, M. (2021) 'Protests in Hong Kong during the COVID-19 pandemic', *Crime, Media, Culture*, 17(1): 17–20.

Jämte, J. and Ellefsen, R. (2020) 'The consequences of soft repression', *Mobilization*, 25(3): 383–404.

Johnston, G. and Johnston, M.S. (2017) '"We fight for all living things": countering misconceptions about the radical animal liberation movement', *Social Movement Studies*, 16(6): 735–51.

Johnston, G. and Johnston, M.S. (2020) '"Until every cage is empty": frames of justice in the radical animal liberation movement', *Contemporary Justice Review*, 23(4): 563–80.

Kahl, A. (2019) *Analyzing Affective Societies: Methods and Methodologies*, London: Routledge.

Katz, J. (1988) *Seductions of Crime: Moral and Sensual Attractions in Doing Evil*, New York, NY: Basic Books.

Kiely, E. and Swirak, K. (2022) *The Criminalisation of Social Policy in Neoliberal Societies: Crime in Late Neoliberal Austerity*, Bristol: Policy Press.

Kindynis, T. (2018) 'Bomb alert: graffiti writing and urban space in London', *The British Journal of Criminology*, 58(3): 511–28.

Kindynis, T. (2019) 'Excavating ghosts: urban exploration as graffiti archaeology', *Crime, Media, Culture*, 15(1): 25–4.

Kindynis, T. (2021) 'Persuasion architectures: consumer spaces, affective engineering and (criminal) harm', *Theoretical Criminology*, 25(4): 619–38.

Kramer, R.C. (2016) 'State crime, the prophetic voice and public criminology activism', *Critical Criminology*, 24(4): 519–32.

Kurmelovs, R. (2021) 'Environmental activists face "fever pitch" of repression from Australian governments, report says', *The Guardian*, 25 November. Available from: www.theguardian.com/ environment/2021/nov/25/environmental-activists-face-fever-pitch-of-repression-from-australian-governments-report-says (accessed 21 July 2022).

REFERENCES

Lee, M. (2021) 'Policing the pedal rebels: a case study of environmental activism under COVID-19', *International Journal for Crime, Justice and Social Democracy*, 10(2): 156–68.

Lefebvre, H. (1991) *The Production of Space*, Oxford: Blackwell.

Little, C. (2015) 'The "mosquito" and the transformation of British public space', *Journal of Youth Studies*, 18(2): 167–82.

Lubbers, E. (2012) *Secret Manoeuvres in the Dark: Corporate and Police Spying on Activists*, London: Pluto Press.

Lundberg, K. (2022) 'Walking at the edges of green criminology: the edges of the city and the extraordinary consequences of ordinary harms', *Criminological Encounters*, 5(1). Available from: http://doi.org/10.26395/CE22050103 (accessed 24 February 2023).

Lundsteen, M. and Fernández González, M. (2021) 'Zero-tolerance in Catalonia: policing the other in public space', *Critical Criminology*, 29: 837–52.

Maguire, H. (2021) 'COVID-19 protests: Vienna considers daytime ban on demonstrations', *The Local*, 16 December. Available from: www.thelocal.at/20211216/covid-19-protests-vienna-considers-daytime-ban-on-demonstrations/ (accessed 20 June 2022).

Manning, P.K. (2003) *Policing Contingencies*, Chicago, IL: University of Chicago Press.

Maroto, M. (2017) 'Punitive decriminalisation? The repression of political dissent through administrative law and nuisance ordinances in Spain', in N. Peršak (ed) *Regulation and Social Control of Incivilities*, London: Routledge, pp 69–88.

Maroto, M., González-Sánchez, I. and Brandariz, J.A. (2019) 'Editors' introduction: policing the protest cycle of the 2010s', *Social Justice*, 46(2–3): 1–27.

Martin, G. (2021) 'COVID cops: a recent history of pandemic policing during the coronavirus crisis', in V. Vegh Weiss (ed) *Criminalization of Activism: Historical, Present and Future Perspectives*, London: Routledge, pp 216–31.

Martin, G. (2022) 'A law unto themselves: on the relatively autonomous operation of protest policing during the COVID-19 pandemic', *Justice, Power and Resistance*, 5(1–2): 28–45.

Marx, G.T. (1998) 'Some reflections on the democratic policing of demonstrations', in D. della Porta and H. Reiter (eds) *Policing Protest: The Control of Mass Demonstrations in Western Democracies*, Minneapolis, MN: University of Minnesota Press, pp 253–69.

Mason, R., Mohdin, A. and Sinmaz, E. (2023) 'Police in England and Wales to get new powers to shut down protests before disruption begins', *The Guardian*, 15 January. Available from: www.theguardian.com/world/2023/jan/15/police-to-get-new-powers-to-shut-down-protests-before-disruption-begins (accessed 16 January 2023).

McClanahan, B. and South, N. (2020) '"All knowledge begins with the senses": towards a sensory criminology', *The British Journal of Criminology*, 60(1): 3–23.

McGovern, A. (2019) *Craftivism and Yarn Bombing: A Criminological Exploration*, Cham: Palgrave Macmillan.

Measham, F. and Brain, K. (2005) '"Binge" drinking, British alcohol policy and the new culture of intoxication', *Crime, Media, Culture*, 1(3): 262–83.

Millie, A. (2011) 'Value judgments and criminalization', *The British Journal of Criminology*, 51(2): 278–95.

Millie, A. (2017) 'Urban interventionism as a challenge to aesthetic order: towards an aesthetic criminology', *Crime, Media, Culture*, 13(1): 3–20.

Millie, A. (2019) 'Crimes of the senses: yarn bombing and aesthetic criminology', *The British Journal of Criminology*, 59(6): 1269–87.

Millie, A. (2022) 'Guerrilla gardening as normalised law-breaking: challenges to land ownership and aesthetic order', *Crime, Media, Culture*. Available from: https://doi.org/10.1177/17416590221088792 (accessed 24 February 2023).

Morato, N.R. (2021) 'Colombia's murderous democracy pre- and post-COVID-19: the assassination of social leaders and the criminalisation of protest', in V. Vegh Weiss (ed) *Criminalization of Activism: Historical, Present and Future Perspectives*, London: Routledge, pp 170–9.

Mudu, P. (2004) 'Resisting and challenging neoliberalism: the development of Italian social centers', *Antipode*, 36(5): 917–41.

Muncie, E. (2020) '"Peaceful protesters" and "dangerous criminals": the framing and reframing of anti-fracking activists in the UK', *Social Movement Studies*, 19(4): 464–81.

Naegler, L. (2012) *Gentrification and Resistance: Cultural Criminology, Control, and the Commodification of Urban Protest in Hamburg* (Vol 50), Berlin: LIT Verlag Münster.

Naegler, L. (2021) 'Resistance and the radical imagination: a reflection on the role of the critical criminologist in social movements', *Critical Criminology*, 30: 225–35.

Natali, L. (2019) 'Visually exploring social perceptions of environmental harm in global urban contexts', *Current Sociology*, 67(5): 650–68.

Natali, L. and de Nardin Budó, M. (2019) 'A sensory and visual approach for comprehending environmental victimization by the asbestos industry in Casale Monferrato', *European Journal of Criminology*, 16(6): 708–27.

Natali, L. and McClanahan, B. (2017) 'Perceiving and communicating environmental contamination and change: towards a green cultural criminology with images', *Critical Criminology*, 25(2): 199–214.

Natali, L., Acito, G., Mutti, C. and Anzoise, V. (2021) 'A visual and sensory participatory methodology to explore social perceptions: a case study of the San Vittore Prison in Milan, Italy', *Critical Criminology*, 29(4): 783–800.

Neville, L. and Sanders-McDonagh, E. (2019) 'Walk this way: the impact of mobile interviews on sensitive research with street-based sex workers', *Tijdschrift over Cultuur & Criminaliteit*. Available from: www.bjutijdschriften.nl/tijdschrift/tcc/2019/3/TCC_2 211-9507_2019_009_003_004 (accessed 5 January 2023).

Oliver, P. and Urda, J.C. (2019) 'The repression of protest in Spain after 15-M: the development of the gag law', *Social Justice*, 46(2–3): 75–99.

O'Neill, M. and Roberts, B. (2019) *Walking Methods: Research on the Move*, London: Routledge.

Ozymy, J., Jarrell, M.L. and Bradshaw, E.A. (2020) 'How criminologists can help victims of green crimes through scholarship and activism', in A. Brisman and N. South (eds) *Routledge International Handbook of Green Criminology* (2nd edn), London and New York, NY: Routledge, pp 150–64.

Pali, B. (2022) 'A criminology of dis/obedience?', *Critical Criminology*. Available from: https://doi.org/10.1007/s10612-022-09664-7 (accessed 24 February 2023).

Pali, B. and Schuilenburg, M. (2020) 'Fear and fantasy in the smart city', *Critical Criminology*, 28(4): 775–88.

Pali, B., Cruz Correia, M.C., Calmet, M., Jones, V., Vranken, L., Mendes, M., Nowak, E. and Požlep, M. (2022) 'The art of repair: bridging artistic and restorative responses to environmental harm and ecocide', in B. Pali, M. Fosyth and F. Tepper (eds) *The Palgrave Handbook of Environmental Restorative Justice*, Cham: Palgrave Macmillan, pp 385–419.

Passavant, P.A. (2021) 'Between crime and war: the security model of protest policing', in V. Vegh Weiss (ed) *Criminalization of Activism: Historical, Present and Future Perspectives*, London: Routledge, pp 103–14.

Peršak, N. (2017a) *Regulation and Social Control of Incivilities*, London and New York, NY: Routledge.

Peršak, N. (2017b) 'Criminalising through the back door: normative grounds and social accounts of the incivilities regulation', in N. Peršak (ed) *Regulation and Social Control of Incivilities*, London and New York, NY: Routledge, pp 13–34.

Peršak, N. (2021) 'Offending sights and urban governance: expectations of city aesthetics and spatial responses to the unsightly', in N. Peršak and A. Di Ronco (eds) *Harm and Disorder in the Urban Space: Social Control, Sense and Sensibility*, London: Routledge, pp 52–77.

Peršak, N. and Di Ronco, A. (2018) 'Urban space and the social control of incivilities: perceptions of space influencing the regulation of anti-social behaviour', *Crime, Law & Social Change*, 69(3): 329–47.

Peršak, N. and Di Ronco, A. (eds) (2021) *Harm and Disorder in the Urban Space: Social Control, Sense and Sensibility*, London: Routledge.

Pink, S. (2015) *Doing Sensory Ethnography*, 2nd ed., London: Sage.

Pleysier, S. (2017) 'Normalisation of behaviour in public space: the construction and control of "public nuisance" in Belgium', in N. Peršak (ed) *Regulation and Social Control of Incivilities*, London: Routledge, pp 95–107.

REFERENCES

PNRR (Piano Nazionale di Ripresa e Resilienza) (2021) 'Piano Nazionale di Ripresa e Resilienza #nextgenerationItalia'. Available from: www.governo.it/sites/governo.it/files/PNRR.pdf (accessed 21 June 2022).

Podoletz, L. (2017) 'Tackling homelessness through criminalisation: the case of Hungary', in N. Peršak (ed) *Regulation and Social Control of Incivilities*, London: Routledge, pp 75–92.

Pontalti, L. (2021) 'Trento, commercianti contro i no green pass: stop ai cortei del sabato nel centro storico', *L'Adige*, 6 November. Available from: www.ladige.it/cronaca/2021/11/06/trento-commercianti-contro-i-no-green-pass-stop-ai-cortei-del-sabato-nel-centro-storico-1.3047796 (accessed 10 November 2021).

Popovski, H. and Young, A (2022) 'Small things in everyday places: homelessness, dissent and affordances in public space', *The British Journal of Criminology*. Available from: https://doi.org/10.1093/bjc/azac053 (accessed 24 February 2023).

Prison Break Project (2017) *Costruire Evasioni. Sguardi e Saperi Contro il Diritto Penale del Nemico*, Lecce: Edizioni Bepress.

Rauhala, E. and Aries, Q. (2022) 'Inspired by Canadian truckers, Europe's "Freedom Convoy" heads to Brussels', *The Washington Post*, 10 February. Available from: www.washingtonpost.com/world/2022/02/10/europe-ban-freedom-convoy/ (accessed 8 July 2022).

Riedel, F. (2019) 'Atmosphere', in J. Slaby and C. von Scheve (eds) *Affective Societies: Key Concepts*, London and New York, NY: Routledge, pp 85–95.

Rius Ulldemolins, J. (2014) 'Culture and authenticity in urban regeneration processes: place branding in central Barcelona', *Urban Studies*, 51(14): 3026–45.

Ruggiero, V. (2003) 'Fear and change in the city', *City*, 7(1): 45–55.

Ruggiero, V. (2021a) *Critical Criminology Today: Counter-Hegemonic Essays*. London: Routledge.

Ruggiero, V. (2021b) 'Concepts for the revitalisation of critical criminology', *The Howard Journal of Crime and Justice*, 60(3): 290–303.

Ruggiero, V. and South, N. (2013) 'Toxic state-corporate crimes, neo-liberalism and green criminology: the hazards and legacies of the oil, chemical and mineral industries', *International Journal for Crime, Justice and Social Democracy*, 2(2): 12–26.

Salter, C. (2011) 'Activism as terrorism: the green scare, radical environmentalism and governmentality', *Anarchist Developments in Cultural Studies*, 1(Ten Years after 9/11: An Anarchist Evaluation): 211–38. Available from: https://journals.uvic.ca/index.php/adcs/article/view/17126 (accessed 8 July 2022).

Scheidel, A., Del Bene, D., Liu, J., Navas, G., Mingorría, S., Demaria, F., Avila, S., Roy, B., Ertör, I., Temper, L. and Martínez-Alier, J. (2020) 'Environmental conflicts and defenders: a global overview', *Global Environmental Change*, 63: 102104. Available from: https://doi.org/10.1016/j.gloenvcha.2020.102104 (accessed 24 February 2023).

Schlembach, R. (2018) 'Undercover policing and the spectre of "domestic extremism": the covert surveillance of environmental activism in Britain', *Social Movement Studies*, 17(5): 491–506.

Schuilenburg, M. and Peeters, R. (2018) 'Smart cities and the architecture of security: pastoral power and the scripted design of public space', *City, Territory and Architecture*, 5(1): 1–9.

Seal, L. and O'Neill, M. (2021) *Imaginative Criminology: Of Spaces Past, Present and Future*, Bristol: Bristol University Press.

Selmini, R. (2020) *Dalla Sicurezza Urbana al Controllo del Dissenso Politico: Una storia del Diritto Amministrativo punitivo*, Rome: Carocci editore.

Selmini, R. and Crawford, T.A. (2017) 'The renaissance of administrative orders and the changing face of urban social control', *European Journal on Criminal Policy and Research*, 23(1): 1–7.

Selva, A. (2021) 'Orsi in Trentino, il presidente Fugatti sotto scorta per le minacce degli animalisti', *La Repubblica*, 25 May. Available from: www.repubblica.it/cronaca/2021/05/25/news/orsi_in_trentino_il_presidente_fugatti_sotto_scorta_per_le_minacce_degli_animalisti-302682748/ (accessed 8 July 2022).

REFERENCES

Sendra, P. and Sennett, R. (2020) *Designing Disorder: Experiments and Disruptions in the City*, London and New York, NY: Verso Books.

Shalhoub-Kevorkian, N. (2017) 'The occupation of the senses: the prosthetic and aesthetic of state terror', *British Journal of Criminology*, 57: 1279–300.

Shennan, R. (2021) 'Controversial policing bill has been delayed after backlash – what you need to know', *The Scotsman*, 19 March. Available from: www.scotsman.com/read-this/controversial-policing-bill-has-been-delayed-after-backlash-what-you-need-to-know-3171419 (accessed 8 July 2022).

Siddique, H. (2021) 'Civil liberties groups call police plans for demos an "assault" on right to protest', *The Guardian*, 11 March. Available from: www.theguardian.com/law/2021/mar/11/civil-liberties-Groups-call-police-plans-for-demos-an-assault-on-Right-to-protest (accessed 11 June 2022).

Simester, A.P. and von Hirsch, A. (2006) 'Regulating offensive conduct through two-step prohibitions', in A.P. Simester and A. von Hirsch (eds) *Incivilities: Regulating Offensive Behaviour*, Oxford: Hart Publishing, pp 173–94.

Smith, N. (1996) *The New Urban Frontier: Gentrification and the Revanchist City*, London and New York, NY: Routledge.

Smith, N. and Low, S. (2006) 'Introduction: the imperative of public space', in S. Low and N. Smith (eds) *The Politics of Public Space*, New York, NY, and London: Routledge, pp 1–16.

Sollund, R. (2017) 'Doing green, critical criminology with an auto-ethnographic, feminist approach', *Critical Criminology*, 25(2): 245–60.

South, N. (2014) 'Green criminology: reflections, connections, horizons', *International Journal for Crime, Justice and Social Democracy*, 3(2): 5–20.

Squires, P. (2008) *ASBO Nation: The Criminalisation of Nuisance*, Bristol: The Policy Press.

Squires, P. and Stephen, D. (2010) 'Pre-crime and precautionary criminalisation', *Criminal Justice Matters*, 81(1): 28–9.

Stephens-Griffin, N. (2022) 'Biting back: a green-cultural criminology of animal liberation struggle as constructed through online communiques', *Crime, Media, Culture*. Available from: https://doi.org/10.1177/17416590221110118 (accessed 24 February 2023).

Stodulka, T., Dinkelaker, S. and Thajib, F. (2019) 'Fieldwork, ethnography and the empirical affect montage', in A. Kahl (ed) *Analyzing Affective Societies: Methods and Methodologies*, London: Routledge, pp 279–95.

Stretesky, P.B. and Lynch, M.J. (2014) *Exploring Green Criminology: Toward a Green Criminological Revolution*, Farnham: Ashgate.

Sumartojo, S. and Pink, S. (2018) *Atmospheres and the Experiential World: Theory and Methods*, London and New York, NY: Routledge.

Szalai, A. (2021) 'A social control perspective for the study of environmental harm and resistance', in V. Vegh Weiss (ed) *Criminalization of Activism: Historical, Present and Future Perspectives*, London: Routledge, pp 30–41.

Taylor, M. and Gayle, D. (2019) 'Thousands block roads in Extinction Rebellion protests across London', *The Guardian*, 15 April. Available from: www.theguardian.com/environment/2019/apr/15/thousands-expected-in-london-for-extinction-rebellion-protest (accessed 12 June 2022).

TGR Trento (2019) 'Processo anarchici. Assolti per l'accusa di terrorismo ed eversione', 5 December. Available from: www.rainews.it/tgr/trento/articoli/2019/12/tnt-anarchici-processo-assolti-reato-terrorismo-e-eversione-73c9c3b8-2319-410b-8792-7fe6d183f436.html (accessed 8 July 2022).

TheLocal.it (2021) 'Italy cracks down on Covid green pass protests', 10 November. Available from: www.thelocal.it/20211110/italy-cracks-down-on-covid-green-pass-protests/ (accessed 14 July 2022).

Thrift, N. (2008) *Non-representational Theory: Space, Politics, Affect*, London and New York, NY: Routledge.

Toledo, I.C., Cavalcanti, R. and Souza, G.I. (2021) 'An analysis of the criminalisation of socio-environmental activism and resistance in contemporary Latin America', in V. Vegh Weiss (ed) *Criminalization of Activism: Historical, Present and Future Perspectives*, London: Routledge, pp 191–200.

REFERENCES

Townsend, M. (2022) 'Government to unveil crackdown on climate activists and strike action', *The Guardian*, 16 October. Available from: www.theguardian.com/world/2022/oct/16/crackdown-strike-action-activism-protests-london-glue-police (accessed 1 November 2022).

Tuzza, S. (2021) *Il Dito e la Luna: Ordine Pubblico tra Polizia e Potere Politico, Un Caso di Studio*, Milan: Meltemi.

Ugwudike, P. (2015) *An Introduction to Critical Criminology*, Bristol: Policy Press.

UK Parliament (2021) 'Police, Crime, Sentencing and Courts Bill. Explanatory notes'. Available from: https://publications.parliament.uk/pa/bills/cbill/58-01/0268/en/200268en.pdf (accessed 24 June 2022).

UK Parliament (2022a) 'Police, Crime, Sentencing and Courts Bill completes passage through Parliament'. Available from: www.parliament.uk/business/news/2021/september-2021/lords-debates-police-crime-sentencing-and-courts-bill-at-second-reading/ (accessed 24 June 2022).

UK Parliament (2022b) 'Police, Crime, Sentencing and Courts Act 2022. Government Bill'. Available from: https://bills.parliament.uk/bills/2839 (accessed 24 June 2022).

Vegh Weiss, V. (2021a) *Criminalization of Activism: Historical, Present and Future Perspectives*, London: Routledge.

Vegh Weiss, V. (2021b) 'Introduction', in V. Vegh Weiss (ed) *Criminalization of Activism: Historical, Present and Future Perspectives*, London: Routledge, pp 1–16.

Villacampa, C. (2017) 'Municipal ordinances and street prostitution in Spain', *European Journal on Criminal Policy and Research*, 23(1): 41–57.

Waddington, D. (2007) *Policing Public Disorder: Theory and Practice*, Cullompton: Willan Publishing.

Walby, K. and Monaghan, J. (2011) 'Private eyes and public order: policing and surveillance in the suppression of animal rights activists in Canada', *Social Movement Studies*, 10(1): 21–37.

White, R. (2013) 'The conceptual contours of green criminology', in R. Walters, D. Solomon Westerhuis and T. Wyatt (eds) *Emerging Issues in Green Criminology: Exploring Power, Justice and Harm*, Cham: Palgrave Macmillan, pp 17–33.

White, R. (2014) *Environmental Harm: An Eco-Justice Perspective*, Bristol: Bristol University Press.

WHO (World Health Organization) (2022) 'WHO coronavirus (COVID-19) dashboard'. Available from: https://covid19.who.int/ (accessed 20 June 2022).

Whyte, D. (2016) 'It's common sense, stupid! Corporate crime and techniques of neutralization in the automobile industry', *Crime, Law and Social Change*, 66(2): 165–81.

Wilson, J.Q. and Kelling, G. (1982) 'The police and neighborhood safety: broken windows', *Atlantic Monthly*, 127: 29–38.

Winlow, S., Hall, S., Treadwell, J. and Briggs, D. (2015) *Riots and Political Protest: Notes from the Post-political Present*, Abingdon: Routledge.

Wood, L. (2014) *Crisis and Control: The Militarization of Protest Policing*, New York, NY: Pluto Press.

Young, A. (2014) 'From object to encounter: aesthetic politics and visual criminology', *Theoretical Criminology*, 18(2): 159–75.

Young, A. (2019) 'Japanese atmospheres of criminal justice', *The British Journal of Criminology*, 59(4): 765–79.

Young, A. (2021) 'The limits of the city: atmospheres of lockdown', *The British Journal of Criminology*, 61(4): 985–1004.

Zedner, L. (2007) 'Pre-crime and post-criminology?', *Theoretical Criminology*, 11(2): 261–81.

Ziniti, A. (2021) 'Il Viminale frena i No Pass. "Basta cortei in centro, sit-in solo con mascherina"', *La Repubblica*, 10 November. Available from: www.repubblica.it/cronaca/2021/11/10/news/il_viminale_frena_i_no_pass_basta_cortei_in_centro_sit-in_solo_con_mascherina_-325757468/ (accessed 10 November 2021).

Index

A

administrative measures/fines 2, 18, 22, 23, 43, 45, 60, 63–64, 102; *see also* ban, incivility regulation
affect 13, 48–49, 52, 55, 57, 71–82, 85–86, 89, 91, 93–94, 95, 97, 102; *see also* atmosphere, emotions, senses
anthropology 79
anti-social (uncivil) behaviour 48, 50, 58, 59–61, 64; *see also* disorder, nuisance
ASBOs (Anti-Social Behaviour Orders) 60–62
atmosphere
  affective 6, 14, 50–51, 72–74, 76–82, 85–87, 89, 90, 93–99, 104
  criminology of 96
  of control 51, 72, 96
  of resistance 14, 71–73, 81–83, 90, 94–99
  ruptured 72, 76, 81, 86–90
  see also affect, emotions, senses
architecture (general) 47, 50, 57, 78
  hostile/defensive 13, 48, 57, 58
  persuasion 49
arts 104
Australia 2, 24, 29
autoethnography 4, 14, 24, 29, 73, 80–81, 89, 93–94, 96, 98; *see also* ethnography, ethnographic place

B

ban 2, 11, 22–23, 38, 42, 48, 60–63, 67–68; *see also* administrative measures/fines, DASPO, PSPOs
barrier 31–*32*, 35–36, 39, 83

bear 8, 31, 39, 90–93, 95
Black Lives Matter 66
Brandariz, J.A. 16, 18, 23, 30, 41, 43, 45, 101–102
Brisman, A. 5, 59, 85
'broken windows' 59

C

Carabinieri 25, 33
Casteller prison 8, 31, 39
city
  centre 4, 6, 8, 13–14, 29, 31, 37–45, 47–49, 62–63, 67–70, 81, 83, 100–102
  inner- 13, 19, 42–43, 45, 47–51, 55, 60, 63–65, 67, 69
  smart 50
  suburb 8, 32
Cohen, S. 103
climate change 2, 3, 6, 99
commodification 55
  of resistance 55
conservation 98, 104
consumption 13–14, 43, 47–53, 55, 57–59, 67, 69–70, 78, 102
containment measures 20, 31, 40, 44, 100; *see also* barrier, militarization, police
COVID-19 (general) 28, 37
  lockdown 3, 69–70, 76
  pandemic 1–3, 6, 12, 19, 23–24, 28, 41, 44, 48, 64, 67, 69, 100–101
  regulations 4, 24, 27, 29, 36–37, 41, 67–68, 87
  vaccines 1, 37, 42
COVID-pass 37–38, 42, 44, 68
craftivism 54
criminology
  activist critical 104

critical 5, 13, 17, 19, 24, 48, 52–53, 74–75, 82, 96, 101–105
critical sensory 4–5, 6, 14, 72–73, 75, 77, 98–99, 102
cultural 5, 9, 48, 52–55, 74–76, 82, 96
green critical 5, 6, 14–15, 101–102, 104
of atmospheres, *see* atmosphere
visual 74–76
criminal law 23, 43–45, 53, 60–62, 64, 102
criminalization 5, 17–19, 22, 35, 41, 44, 46, 48, 53–55, 60–61, 103; *see also* penalization, repression
container space 43
cultural space 52–53
curfew 36, 41, 60

D

DASPO (Italian ban on accessing urban areas) 62–63
depoliticize 16, 30, 40, 44, 55
DIGOS (General Investigations and Special Operations Division) 25, 27, 29, 33, 37, 40, 86–88
disobedience 103, 105
disorder 6, 13, 27, 45, 47, 50, 59; *see also* anti-social (uncivil) behaviour, nuisance
displacement 13–14, 37, 42, 44–45, 48, 56–57, 68, 100–101; *see also* gentrification
dissent 13, 19–20, 22–23, 29, 35–36, 43, 56, 69, 103
DPCMs (Decrees of the Italian Prime Minister) 11, 24

E

eco–justice, *see* justice
emotions 5, 57, 74–77, 79; *see also* affect, atmosphere, senses
enemy 18, 40, 46, 100
England 3, 60, 62–63, 65, 67, 70

ethnography 10, 12, 19, 26, 81, 100; *see also* autoethnography
ethnographic place 81–82, 94–95
Europe 1–2, 7–8, 14, 47–48, 59–60, 62–65, 67–70, 100
exclusion 13, 35, 42–43, 46, 50, 52, 55, 58, 63, 69, 102
Extinction Rebellion (XR) 2, 24, 29, 65–67, 69
extremism 17

F

Ferrell, J. 10, 52–53, 103
fieldnotes 26, 73, 81–83, 85–86, 88–90, 94
fieldwork 26–27, 29, 36, 38–40, 68, 73, 79, 81–82, 93–94, 98, 101
flag 71, 83, 87–88, 91–93, 101
flower 78, 90–91, 93, 97
flyer 3–4, 23, 26, 30, 39, 64, 71, 83, 88
Foucault, M. 50–51
fracking 21, 66
Fraser, A. 72, 77–79, 81, 89, 95–96

G

Garcia Ruiz, A. 33, 49, 58–59, 77
gentrification 13, 55–57, 64; *see also* displacement, regeneration
geography 48, 54, 75
Global North 17
Global South 17
González-Sánchez, I. 16, 18, 23, 30, 41, 43, 45, 101–102
governance 4, 12, 27, 100, 102
green critical criminology, *see* criminology
graffiti 3–4, 9, 53–54, 71, 78, 101
Guardia di Finanza 25, 33
guerrilla gardening 54, 76

H

HS2 (High Speed 2) 66
Hayward, K.J. 9, 43, 47, 49–50, 58–59, 74–75

harm 5–6, 9, 14, 17, 29, 37, 49–50, 57, 59, 71, 73–74, 77, 82, 86, 89, 91, 94–99, 102–105
health crisis 3–4, *see also* COVID-19
homeless 13, 57–59, 62, 64
humanities (academic disciplines) 6, 104

I

image 4, 75–76, 90
imagination 4, 54, 70, 73, 80–82, 93–96, 98–99, 102–105
incivility regulation 13, 48, 56, 59–60, 62–64; *see also* administrative measures/fines, ban
Insulate Britain 67
interdisciplinary 5–6, 14, 69, 96–99, 104
intersectionality 9, 103
Italy 8, 10–12, 19–25, 27–28, 31, 38, 42–44, 62–63, 70, 100

J

Just Stop Oil 67
justice
  criminal 61, 72, 76
  eco- 1–21, 24–25, 29, 31, 37–39, 41–47, 68–73, 85, 93, 95, 97–102
  environmental 5, 9, 93–94
  ecological 3, 9
  species 5, 9, 93–94

K

Kindynis, T. 49, 54, 78–79
Kramer, R.C. 103–105

L

labelling 16, 20, 40, 42, 46
Lefebvre, H. 52, 54
lockdown, *see* COVID-19

M

Maroto, M. 16, 18, 23, 30, 41, 43, 45, 64, 101–102

Martin, G. 18, 24, 40, 45, 102
mask/facial coverings 2, 12, 23, 25; *see also* COVID-19
McClanahan, B. 5, 14, 27, 33, 49, 72, 75, 77, 96, 102
megaproject 21, 35, 86, 91, 98
military 38
militarization 18, 21, 27, 31, 35, 39–40, 102
Millie, A. 52–54, 76–77, 85, 94, 96
migrants 13, 38, 43, 56, 62, 64
more-than-human 14, 74, 77, 96–99, 102, 104
Mosquito device 58–59
multidisciplinary 4–6, 69, 102
music 54, 58

N

Naegler, L. 9–10, 55, 104
neoliberalism 19, 23, 43, 47, 56
NOTAV (anti-high-speed railway committee) 7, 9, 12, 14, 21–22, 26, 28–32, 35–40, 42, 44, 81, 83, 85, 87, 91–93, 95
NOTAP (anti-Trans-Adriatic Pipeline movement) 21, 35, 40, 44
nudge 13, 48, 50
nuisance 22, 48, 50, 56, 59, 63–65; *see also* anti-social (uncivil) behaviour, disorder

O

O'Neill, M. 82, 96
order
  aesthetic 53, 77, 94, 96
  civil 53, 60–63; *see also* ASBOs
  normative 70, 78
  public 13, 19, 25, 27, 29, 40–41, 44–45, 100
  social 10, 41, 51, 54, 75, 102
  spatial 54

P

Pali, B. 13, 49–50, 97, 103, 105
pandemic, *see* COVID-19

penalization 2, 19, 65, 101–102; *see also* criminalization, social control, repression
Peršak, N. 55–61, 72, 74, 96
Pink, S. 14, 51, 72–73, 79–82, 93–95, 97, 99
PNRR (Piano Nazionale Ripresa e Resilienza or 'National Recovery and Resilience Plan') 7–8, 91
police (general) 2, 4, 7, 8, 13, 18–31, 33–46, 55, 59, 63–68, 70, 86–90, 95, 100–102
  cordon 31, 33; *see also* barriers, containment measures
  riot-gear 20, 27–29, 33, 35, 38–39, 41, 87
  riot-control vehicles 21, 28–29, 33, 35, 38–39, 41–42, 87, 89
  see also militarization, policing, power
policing
  protest 3, 6, 13, 19–21, 28, 30–31, 40, 44–46, 81, 100
  spectacle 41
  'zero-tolerance' 56
  see also militarization, police, power
power
  affective/atmospheric 48–49, 51–52, 75, 78, 89–90, 95
  disciplinary 51, 76, 91, 95, 99
  police 3, 18, 33, 38–40, 42, 44, 61, 65–67, 90
pre-crime 59
prefect 25, 38, 42, 68
private security 18, 35
protest
  disruptive 2, 42–43, 45, 65–69
  policing, *see* policing
  regulation of 1–4, 11, 13–14, 18, 23–25, 36–39, 41–43, 45, 48, 59, 64–70, 87–88, 100–102
  right to 1–2, 13–14, 18, 23–24, 31, 38, 48, 64, 66–67, 69–70
  static 24, 36–38, 41, 69, 87
  transgressive 20–22, 28, 31, 40, 42–45, 100
PSPOs (Public Spaces Protection Orders) 61–62

Q

questore 22, 25, 27, 29, 37, 40, 62–63
questura 25, 37, 41–42, 45, 68–69

R

regeneration 13, 47–52, 55–56, 60, 63–64; *see also* gentrification
repair 97, 98–99, 104
repression 23–24, 26, 30, 35, 70, 99; *see also* criminalization, penalization
resistance (general) 3–4, 6, 8–10, 51, 55, 72, 85, 97
  atmospheres of, *see* atmosphere
  commodification of, *see* commodification
  visual 11, 14, 71, 82, 94–97, 101–105
  performative 71, 73, 94
  practices of 1, 4, 10, 71, 100
right to protest, *see* protest
risk/risk governance 25, 29, 45, 56, 59, 62–63
river (general) 90–91, 93, 97–98
  Adige 6, 81, 90
  Zenne 97
Ruggiero, V. 16–17, 102, 105

S

San Lorenzo bridge 81–82, 90, 92, 93, 97
San Martino neighbourhood 37, 83–*84*, 86
SC (StopCasteller campaign) 7–9, 12, 26, 28–31, 36–40, 42–44, 101
Schuilenburg, M. 9, 13, 43, 49–50, 58, 78
Seal, L. 82, 96
Selmini, R. 22–23, 38, 60, 62, 64
senses (general) 71–72, 74, 76–77, 82, 85–86, 91, 93–94, 98, 102

hearing 58–59, 64, 86, 93, 97
sensory memory 82, 93–95
sight 58, 70, 74, 77, 88–91, 93
smell 50, 58, 91, 93
technologies affecting the 13, 48, 58–59, 64
see also affect, atmosphere, emotions
sense of place 94–95
sex work 13, 38, 43, 56, 62–64
skateboarding 54, 57
social centre 8, 21, 28, 81
social construction 41, 45, 47, 65
social control 1, 3, 4–6, 10, 13–14, 25, 47, 100–101
social media 4, 11, 18, 26, 87, 101
social movement 5, 9–10, 12, 16, 29, 45, 56, 70, 101–102, 105
 animal rights 7–8, 10, 20
 eco-justice 1–12, 16, 19–21, 31, 44, 93, 100–101
 environmental 8, 17, 65–66
 anti-globalization 20–21
 Indigenous 18
 transgressive 21, 28, 42
social change 103
social sciences 6
sociology 48
South, N. 5, 14, 33–34, 49, 58–59, 72, 75, 77, 85, 96, 102
space
 of consumption and pleasure 47
 see also container space, cultural space, sense of place
Spain 23, 64
Squadra Mobile 25, 27
state of exception 3, 18, 22–24, 40, 44–46, 101
sticker 9–10, 71, 90, *92–93*, 101
Sumartojo, S. 14, 51, 72–73, 79–80, 82, 93–95, 97, 99
surveillance 5, 17–18, 21–22, 24, 26–27, 33, 35, 46, 51, 67, 87, 89, 95
sustainability 4, 7, 73–74, 96, 104

T

TAV (Treno Alta Velocità or 'high-speed railway') 8–9, 21–22, 33–35, 37, 83, 85–87, 89, 91, 98
terrorism 16–18, 22
threat 18, 25, 30, 39–40, 50, 63, 65
Thrift, N. 49, 78
Trento 4, 6–8, 10, 14, 19, 26, 28–29, 31–*32*, *33*, 35–46, 68–69, 73, 81, 91, 100–101

U

UK 2–3, 21, 47, 48, 54, 60–61, 63, 65, 67, 69
urban design 13, 47, 49–51, 57–59
urban interventionism 54, 76
urban planning 51–52, 57
urban poor 13, 38, 43, 56, 58
urban studies 13, 48

V

Vegh Weiss, V. 16–17, 101
virus, *see* COVID-19
visibility 4, 13, 27, 29, 40, 44–45, 83, 100
visual criminology, *see* criminology

W

walking 11, 14, 23, 33, 35, 37, 40, 71, 81–85, 87–91, 93, 97–98
White, R. 2, 5, 9, 16–18, 41

Y

yarn bombing 54, 76
Young, A. 23, 51, 72, 75–77, 79, 89, 93–96
young people 1, 3, 13, 52–53, 58–62, 64, 86

Z

'zero-tolerance' policing, *see* policing

Printed and bound by CPI Group (UK) Ltd, Croydon, CR0 4YY

08/11/2023

08185994-0001